GLOBAL MARINE ENVIRONMENT

Does The Water Planet Have A Future?

Herman Prager

UNIVERSITY PRESS OF AMERICA

Lanham • New York • London

University Press of America®, Inc.
4720 Boston Way
Lanham, Maryland 20706

3 Henrietta Street
London WC2E 8LU England

Printed in the United States of America
British Cataloging in Publication Information Available

Library of Congress Cataloging-in-Publication Data

Prager, Herman.
Global marine environment : does the water planet have a future? / by Herman Prager.
p. cm.
Based on the author's dissertation (Ph.D.)—University of New Orleans.
Includes bibliographical references and index.
1. Marine pollution. 2. Oceanography and state. 3. International relations. I. Title.
GC1085.P73 1993 363.73'94—dc20 92–44141 CIP

ISBN 0–8191–9016–0 (cloth : alk. paper)
ISBN 0–8191–9017–9 (pbk. : alk. paper)

The paper used in this publication meets the minimum requirements of American National Standard for Information Sciences—Permanence of Paper for Printed Library Materials, ANSI Z39.48–1984.

This work is dedicated to the memory of my late mother,
Katherine Landry Prager, and to my father,
Herman J. Prager, Jr.

The Earth is one but the world is not.

World Commission on Environment
and Development

The Earth is a planet. We know for sure there are only eight others, all within our solar system. Whether or not in this or future decades other planets are discovered among the tens of billions of stars existing in our galaxy alone, surely it is now the time for all the peoples on the Earth to realize the radical and fundamental unity which life places upon its sentient members.

Acknowledgements

These are difficult to write for several reasons. First, of course, there have been so many people who have helped me over the years that to try and list all of them would be futile indeed. This work began as a doctoral dissertation at the University of New Orleans, and then proceeded into book form while a new faculty member of Northeast Louisiana University. There is also a temptation, perhaps for new authors, to pen something such as "Thanks to everyone who helped" and there let the matter rest. That would hardly be appropriate. Therefore, let me mention some people who contributed especially to the creation and publication of this book while at the same time stating there are many who have been left out (but not for lack of gratitude and sincere thankfulness). Professor Richard Haynes, chair of the Department of History and Government at Northeast Louisiana University, provided continuous encouragement and support in transforming the manuscript into a acceptable copy for the publisher. In addition, a special word of appreciation must be given to Dean Frank Morgan, Jr. of the College of Liberal Arts and Acting Dean H. T. Garner of the Graduate School for the essential financial support their offices provided. Beth Wright, a graduate student in the department of history and government, was an integral assistant in this entire process. Without her extensive knowledge of word-processing skills the current work would have been seriously delayed.

The genesis of the ideas contained within these pages were first developed within the political science department at the University of New Orleans. There, the chair of my doctoral committee, Professor Robert Jordan, was always a wise counselor and a repository of good advice—professional and otherwise—particularly when one or two "bumps" developed along the road. His unfailing approachability, sincere concern and interest, and high professional standards were of inestimable value. I doubt I could have worked so well without him as chairperson of the committee. Other members of that group were equally important in providing a wealth of sound advice and helpful suggestions. I especially am thankful for those who pointed me in the direction of articles or books I had overlooked in my own research. They also provided critical, but very useful, suggestions on how and why to tighten the focus of a section or chapter, or to include the discussion of some theory which was particularly relevant or pertinent to the argument being made. For these specific points, and perhaps more importantly for providing a supportive and genial intellectual atmosphere in which to work, I thank Professors R. Judson Mitchell, Alan Leonhard and Lawrence Katzenstein.

A special word of thanks must be made to Dr. Clive Archer of the Centre for Defence Studies at the University of Aberdeen. Dr. Archer served as the outside reader, and his expertise especially in the areas of ocean management, international organizations, fisheries and the politics of Northern waters was particularly constructive and salutary.

I would also like to thank the following copyright holders for the privilege of quoting from their works. *The Anarchical Society* by Hedley Bull, copyright 1977, published by Columbia University Press, New York. Reprinted with the permission of the publisher. *Global Technopolitics: The International Politics of Technology and Resources* by Dennis Pirages, copyright 1989, published by Brooks-Cole Publishing Company, a division of Wadsworth, Inc. M. Grant Gross, *Oceanography: A View of the Earth*, 4th ed., copyright 1987, pp. 32, 163, 164, 176, 187, 258, 268, 269, 297, 298, 357, 359, 365 384, 398. Reprinted by permission of Prentice-Hall, Englewood Cliffs, New Jersey. And from *Politics Among Nations*, 3rd ed., by Hans J. Morgenthau, copyright 1948, 1954, (c) 1960 by Alfred A. Knopf, Inc. Reprinted by permission of the publisher.

For many years, starting as a young boy around the age of 12, I was at first fascinated and enthralled as were millions of others, and then impressed by the work of a former French naval officer who was involved in the creation of television documentaries for a world-wide audience, Jacques-Yves Cousteau. Continuing with this interest and at the same time accompanying my father and mother on deep-sea fishing trips into the Gulf of Mexico (mostly, it seems, to read and look for whales and dolphins), the ground work was laid, I believe, for a subsequent professional interest in issues of ocean degradation. Beginning in the mid-1980s, I also began to notice in press articles and in television programs the statements and speeches on environmental matters which were then being made by The Prince of Wales. His Royal Highness has continued to challenge governments and individuals both to reevaluate and alter the dangerous, and ultimately untenable, industrial and philosophical habits which are endangering this unique creation of the Solar System on which we live. In a very broad sense, this book and the concern for the future of our planetary environment which motivated its research and writing, was first encouraged by the life work of these two very different individuals, of quite different stations, who are making significant contributions to a deeper global understanding of the importance of environmental problems.

On a personal plane, I owe a particular debt of thanks to three very different people, without whom I would not have been able to complete the long journey that is sometimes represented by the doctoral degree. Each in their own way has been a special and unique friend: my grandmother, Anna R. Prager, Rev. Ellis DePriest, S. M., and Paul M. Sanchez-Navarro.

Naturally, all omissions or errors are the sole responsibility of the author.

Contents

Preface

Justification

The modern international system that we know today, comprising approximately 170 self-proclaimed nation-states, is in the most precarious and weakened position since its inception more than three hundred and forty years ago. The viability of the system has been threatened by a series of long historical processes begun many centuries before our own, and is a serious challenge to the theory of international relations and sovereignty. The signs are everywhere that the planet is in peril and the work of scholars and the attempts to remedy the situation through inter-governmental means appears to be inadequate. An apparent liability has to do with an international system based on state-to-state relations regulated by the notion of "sovereignty."[1] Furthermore, the threat of nuclear conflict, although diminished since the height of the Cold War, is still a sword of Damocles hanging over a world divided by profound and longstanding linguistic, cultural, economic, religious, historical and political differences. Finally, environmental destruction and the attendant threat to species survival (including our own: homo sapiens sapiens) has now assumed widespread and genuine concern among all levels in government and society in the West. The so-called underdeveloped world has become aware of the problem as well, even as they persist in their attempts to industrialize and modernize.[2]

For reasons of methodology, however, this study will focus only on the state system and the problem of ocean degradation as practiced in the West; specifically the Western industrial market economies. This on-going ecological debasement is the most serious threat to human survival because it is endangering the very balance of life itself. These are the complex threads and interactions that make biological and chemical processes exist and which are essential to the creation and the sustaining of the life characterized by growth, reproduction, metabolism and response to stimuli that are now present on Planet Earth. At the heart of these controversies lies the core of what many have said the study of political science is all about—the allocation of power and values in society. In other words, it is about the adjudication between differing public policies in response to divergent appreciations on the nature of social challenges and appropriate responses.

This debasing of the environment is nowhere more critical than in the many bodies of seas and oceans that are more accurately and efficaciously called the Global Ocean. Covering about 71% of the surface of the earth,[3] the water planet is a living organism[4] which reacts both to the encroachments made upon it and to the natural processes occurring within its vast expanse not contingent upon human intrusion or application.

The current international system since its inception has not had to address, in any significant manner, ecological problems.[5] This is because the perception of the necessity had not previously arisen although the economic and social—as well as environmental—consequences of industrialization, since the 19th century onward, has had profound consequences.[6] The Western understanding of what constitutes "the good life" has had profound effects in all areas of human life and in the life of states. The changes in the human condition brought about by accelerating industrialization—particularly in the 20th century—and attendant structural alterations in social organization have created a radical deterioration in the ecology of the planet.[7] The deterioration threatens the very basis of what has come to be called post-modern civilization. During this period of transformation the politics among states have undergone an inversion. Historically, states have been concerned almost solely with levels of power expressed in strictly military terms. Is there a "balance?" What would be the results for the states involved if the "balance" were threatened?[8] Questions, including the economic expansion of European states into overseas territories previously undiscovered, were also directly related to these strategies. It might be useful to add, parenthetically, that these news lands were only "undiscovered" to the Europeans. The areas were certainly "discovered" long before by the native inhabitants living there, and in many cases great civilizations were built (particularly the Aztecs and Incas in what has come to be called Latin America).

Historically, the notion of "power"—its definition and use—has been the dominant factor in the relations between states. Written records of society have established this since the dawn of the human species. In only approximately the last ninety years, however, the components of this power, i.e. the factors that go to make it up and which animate its exercise, have greatly expanded. The question of what is a threat now to the survival of a state, in the late twentieth century, is broader than simply the organized use of violence. This point has been developed more fully in the writings of such political economists as Mancur Olson.[9]

There are many tools available for influence or undermining the power or status of other nation-states. Economic variables (trade and investment patterns, and embargoes), scientific and technological issues (especially involving technology transfer) and environmental dangers (including nuclear and other hazardous waste) can all prove as salient a threat to national security in the 1990s as outright military force did in decades and centuries past. These three tools or factors are rearranging previous understandings of what constitutes national security and power.

Interdependencies in economic affairs (trade, loans, balance of payments, inflation, exchange rates, debt payments and competitiveness) have become tightly interwoven since the end of the Second World War. The scope for freedom of involvement in the international arena has been drastically circumscribed by these events. Nations cannot act as unilaterally and imperiously, and with the same relative degree of lack of cost, as they once did (say, in the 19th century with the ubiquitous gunboat). Particularly important or powerful nations do not have the option of ignoring traditionally defined "weak states" which in times past could be shunted aside either through threats or outright military action.[10]

As the international order has been transformed so, too, have the issues which confront the international community. Forty-five years ago the following items were not on the agendas of national leaders or foreign offices: the greenhouse effect, nuclear waste disposal and dumping, hazardous materials transportation, acquired-immune-deficiency syndrome (AIDS), desertification, biodiversity, drift-netting, narco-states and drugs, or ozone depletion. In 1968 the international community, operating through the United Nations system, decided to issue a call for a general conference on the human environment. This became the 1972 Stockholm conference. It was only after this extraordinary gathering that the state's and people's of the world could point towards a specific international organization designed to begin the process of addressing, through coordination within the United Nations system of agencies and organizations, global environmental questions. Yet this effort, as the story to follow illustrates, is not as far-reaching or influential as one might suspect given the enormity of the problem. That state system is still struggling over the initial questions of adjudicating between different claims and rights in regard to such basic questions as: What is pollution or environmental degradation, who is responsible for it, who should pay for clean-up and restoration efforts, how should these matters be apportioned within the domestic structures of member governments (i.e. what role should international law play vis-a-vis domestic law), and what should be the scientific processes instituted to clean up the identified problems? Perhaps, too, the greatest question is, and it is the one this study will address specifically with regard to the ocean: Can variously cooperating state governments, in part operating through intergovernmental organizations, actually solve the problems?

Of all the planet-wide environmental problems, none is more serious than the degradation of the ocean. Earth is a water planet, and every serious pollution finds its way into the ocean. The problem itself either originates in the ocean, or is carried to it by the seas and gulfs and rivers that eventually empty into it. This occurs primarily through hydrological and meteorological processes that interact with the planet. As the world explorer and oceanographer Jacques-Yves Cousteau has noted, "All pollution is ocean pollution."[11]

This ocean pollution now affects every region on the planet, including once hitherto pristine areas such as the Antarctic. By its very nature, pollution often transcends the borders of any one state. The community of nations is not effectively organized or well prepared to solve the myriad of problems continually thrusting themselves to the forefront of the international agenda. As this research will make clear, the nations of the world have been stymied in their efforts to reach comprehensive solutions. This inability to act decisively is posing dire consequences to the people of the planet—irrespective of national origin, ideology or economic status—in terms of livelihoods, lifestyles, health and the ability of peoples to solve peacefully such essential quality of life questions. The next 30 to 70 years may very easily be ones of general environmental emergency unlike any other civilized humanity has faced.[12]

What are the causes of this emergency? Why has the community of sovereign states not responded aggressively and comprehensively to move against such problems, if indeed they are as serious as stated above? The story of this evolution is of central concern in this book. Another way to express the research question is to ask whether or not the international system, as currently organized, can reduce and rectify the destructive ecological processes now operative within the ocean and what is required to transform the existing system in order to avoid catastrophe?

Chapter Topics

Chapter 1 addresses the evolution of the contemporary state system by examining the concepts of development and progress, and the three great revolutions that have shaped human life. Chapter 2 moves on to address the theoretical aspects of the current state system in light of the contributions of international relations theory as its relates to states–their origin, development, extant problems and proposed solutions. Chapter 3 is an analysis of the physical state of the ocean and the problem of pollution. Part I of that chapter examines basic principles in oceanography which impact upon this issue and the constraints present on the international system from the biological, chemical and physical processes that are operative in, this, the most dominating feature of Planet Earth. Part II subsequently analyzes five major components of the problem of ocean pollution. Next, in Chapter 4, the contributions of four United Nations agencies are discussed in light of their attempts both to address the status of ocean degradation and the efforts being undertaken to ameliorate or reduce the existing causes or symptoms. These intergovernmental organizations (IGOs) operate, of course, in an international environment dominated by territorially based sovereign states. The role and importance of international non-governmental organizations (INGOs) in the protection of cetaceans will be taken up in the following chapter by specifically focusing on the work and activities of The Cousteau Society and Greenpeace. Finally, the conclusion posits the question as to whether the state system can adapt quickly and adequately enough to manage and solve the problems of the global ocean.

Chapters 1 and 3 do not have conclusions. This is done purposely. These chapters function more as reports or straightforward accounts of the topics presented, rather than the more analytical components that characterize the remaining issues examined in this book.

Preface Notes

[1]I define sovereignty as a principle of international law which elevates the concept of non-interference in the domestic affairs of other states as a supreme value. Sovereignty also assumes nation-states will pursue their own self-defined interests within the international system. Finally it holds that, in international law, the individual nation-state alone is the sole arbiter of its actions.

[2]See *The New York Times*, 31 July 1989, p. 1. Fernando Collar de Mello, candidate for president of Brazil's 1989 elections, called for Brazilian acceptance of foreign aid to save the Amazonian rain forests. Of course, the call could have been more a domestic political ploy to shift the costs of saving this ecological resource onto others; but it does not negate the significance of his original statement.

[3]M. Grant Gross, *Oceanography, A View of the Earth*, 4th Ed. (Englwood Cliffs, NJ: Prentice Hall, Inc., 1987), Table A 2-1, p. 365.

[4]James Lovelock, *The Ages of Gaia, A Biography of Our Living Earth* (New York: W. W. Norton & Co., Inc., 1988), passim.

[5]See Anna Bramwell, *Ecology in the 20th Century, A History* (New Haven and London: Yale University Press, 1989), p. 4. She notes: "The word ecology is widely used today in the normative sense, not in the biological sense. The science of ecology is one that considers energy flows within a closed system. The normative sense of the word has come to mean the belief that severe or drastic change within that system, or indeed any change which can damage any specie within it, or that disturbs the system, is seen as wrong. Thus, ecological ideas have come to be associated with the conservation of specific patterns of energy flows."

[6]See, for example, B. D. Gooch, *Europe in the Nineteenth Century* (London: The Macmillan Co., 1970), passim.

[7]This point is discussed further in Chapter 5 in regard to the influence of the work of The Club of Rome on contemporary thinking.

[8]See Thucydides, *History of the Peloponnesian War*, trans. Rex Warner (London: Penguin Books, 1954); Sun Tzu, *The Art of War*, trans. Samuel B. Griffith (New York: Oxford University Press, 1971); and Miyamoto Musashi, *A Book of Five Rings*, trans. Victory Harris (Woodstock, NY: The Overlook Press, 1974).

[9]See the review article by Robert B. Reich "Why Democracy Makes Economic Sense," *The New Republic*, 19 December 1983. This article is a thorough review of the very large issues raised by a study of political economy. Reich specifically directs his attention to a review and integration of Mancur Olson's *The Rise and Decline of Nations* (New Haven: Yale University Press, 1982).

[10]The example of the 1973 Arab-Israel war and the subsequent embargo on oil shipments placed against the United States and other Western States is a case in point. Traditionally defined "weak states" successfully pulled off a veritable "coup d'etat" against the traditional notions of power and influence.

[11]Telephone interview with Neal Shapiro, Environmental Analyst, The Cousteau Society, Los Angeles, CA (13 November 1989).

[12]I owe this observation to the head of a research program at the World Meteorological Organization in Geneva, Switzerland. Personal interview with the author (13 April 1989).

1

The Evolution of the Contemporary State System

The Concept of Development and Progress

The material level of achievement, in the sense of the quantity and quality of goods produced by some nations and societies, in the late 20th century is quite simply staggering when compared to two realities. First, the number of such products available compared to historical societies in centuries past is very significant. Second, the distribution of such goods or products is very uneven over the planet. As an observer in the early 1980s noted, despite global interdependence being a fact of life by at least the last three decades of the current century, humankind still does not fully grasp the exigencies of life in a Global Village. "While a small portion of humanity travels ... [on] first class [accommodations aboard "Spaceship Earth"] ..., the rest are down in steerage."[1] In addition, disparities in incomes and standards of living of those in "first class" accommodations contrast dramatically not only with those in "steerage" but also with those who lived during previous eras in human history. There is a quantum leap in magnitude present. In all indicators of life expectancy—infant mortality, scientific achievement, economic production, and education—the average person living today in the industrialized market economies of the West has a higher percentage chance of availing themselves of the fruits of 20th century society than any others living anywhere else on the planet.

Of course, a similar comparison, suitably modified, could be offered between the poor of today and the majority of individuals who lived during all previous recorded human history. Such comparisons are obvious and self-evident. In many ways the poor, in the West, are considerably better off than their contemporaries elsewhere. Such considerations, though, are not what is at issue here. The point is that the wealth or standards of living for those fortuitous enough to have been born into the "right country" have a much greater probability of avoiding what euphemistically could be described as difficult or painful lives.

In different circumstances, the late American president John F. Kennedy remarked that life was unfair. Certainly such considerations could be shared by men and women everywhere, under varying conditions. Not all humans on the planet, however, face the same issues of distributive justice as a farmer in Africa, a tradesman or woman in South America or a barrio dweller in the Philippines.[2]

This chapter begins by analyzing the problem of just what development has come to mean in human history and what this notion of progress entails. This is undertaken for several reasons. First, the greatest level of material progress, as we have said, has been achieved in the West.[3] Second, processes of deep-rooted and transforming social and economic change have been operative for longer periods of time in the West than in any other part of the globe.[4] I date this process for modern time some 200 years ago, at the beginning of the Industrial Revolution in England. And, third, despite the very real and disruptive components (and destructive as well) of industrialization, this paradigm is still the vehicle of choice for almost all nations of the Third World. It is the desire for "modernization" with all its trappings of consumer goods and high levels of consumption that drive so many societies away from traditional structures. This is so despite the desires of some national elites to modify, eliminate or completely restructure those aspects of modernization they deem undesirable or unacceptable to their traditionally based cultures.[5] Many in the non-industrialized world, of course, have great difficulties with the premises of Western industrialization.[6]

In some ways this conflict is as simple as the demand "what you have, we want." From the formal ending of the colonial era—in the 1950s and 1960s—through the era of strengthened (and temporary) Third World solidarity in the 1970s, into the decade of retrenchment and limited global cooperation in the early 1980s; the underdeveloped global majority have been repeatedly frustrated in their desire for improved living standards and economic growth.[7] In other ways, the needs and desires for such changes are more multifaceted than the simple declaration previously noted. The social, economic and historical problems of the Third World are beyond the scope of this present research. Yet a vast literature exists in the field.[8] The most serious question of global peace is raised by a situation in which the overwhelming majority of the world lives in abject poverty, sub-standard health and housing; with little or no reasonable hope for improvement. Even this does not include the many thousands who each day are literally starving to death somewhere on the planet.

The story of civilization is the story of adaptation, insight, seized and lost opportunities, and of technological and scientific discovery. It is a great drama of human creativity. Its dynamism is only partly accessible to human understanding, as any scholarly work can only tap into part of the complex called reality. Yet, the scholarly work of centuries has yielded very concrete results.

The last 200 or so years have been like no other because the pace and scope of change has been inexorably and rapidly increasing. It is almost as if history itself is occurring more quickly. This acceleration began with what has been described as an

Industrial Revolution in mid-eighteenth century England which harnessed water power to transform production systems. This process began with small scale production, but as the nineteenth and twentieth centuries progressed very large factory assembly became dominant; eventually employing thousands in single industrial locations. Water power became steam power and then electrical power in the late 1800s. Later, refined petroleum products were used to drive industrial processes. Since the end of World War II, nuclear power generating plants have also come to play important roles in the energy output of many nations (particularly in Western Europe, and especially in France). By 1945, in the United States, industrial production assumed massive proportions. Consumer products became widely available and at relatively inexpensive prices. The phenomenon of the assembly-line helped create the massive outpouring of armaments necessary to wage the war, and after the fighting ended were redesigned and expanded to create the largest continuous production of consumer and durable goods the world had ever seen. The non-industrialized or Third World nations shared only fractionally in this economic growth, but enough to begin their own processes of environmental assault.[9]

The basis of this increased material wealth is organized around the concept of exploitation. What is meant by this term is the very simple one that nature is viewed as an inexhaustible resource, infinite (or nearly so); one designed to offer humanity a near unlimited extractive capability—or so it seems.

Differing World Views

Dennis Pirages has written extensively on the underlying bases of this current age. His work has been among the most important published in the political/environmental field. Pirages has analyzed, as part of an overall understanding of environmental threat and challenge, the impact that dominant scientific paradigms exert in contemporary society.[10] The concept of differing world views assumes a prominent place in his thinking. In a 1989 work he observed that

> The atypical four-hundred stretch of recent history dominated by the industrial revolution has given rise to a culture or predominant world view that could best be called *exclusionist*. This world view permeates social life and consciousness as well as academic enterprise in the industrial world. It is manifest in the beliefs that human beings exist apart from nature and that they are destined to dominate it, that the last four hundred years of growth and progress will be followed by a similar stretch of good fortune.[11]

This exclusionist design has gone hand-in-hand with certain developments in Christian theological thinking. The Protestant Reformation in the 16th century of Europe revolutionized Western religious thought and life.[12] Older patterns of authority and influence in many countries of the continent (particularly, but not exclusively in the north) were overturned and replaced. New political relationships supplanted religious and monarchical patterns of five centuries.[13]

The new political relationships also went hand-in-hand with new modes of economic organization and intercourse.[14] Monarchs which once co-existed in a papal condominium and exercised their national and extra-national authority in a shifting environment of cooperation, belligerency, tact and guile, broke the centuries old bonds of these political and religious relationships and began the process of developing the modern nation-state. These kings which once shared their dominion and governance in a purported common Christendom now found their specific domestic political concerns radically changing.

In broad historical outline, the dawn of the Christian era resulted in the establishment of the Roman primacy as the pre-eminent center of all the early Sees. This was true not only in a theological sense, but in a political one as well. The groundwork was laid when the Emperor Constantine in 312 AD adopted the new religion as the official one of the Roman Empire. The civilized Western world became Christian and overwhelmingly Roman Catholic.[15]

This history of church and (proto)state relations precluded a national independence free of intimate papal influence. Such direct involvement with the "court politics" of a religious-cum-political leader is what sparked a little more than a millennium later the Protestant (protest) movement against Roman influence and control over national political life. One of many results of this increasing national monarchical power was the concomitant weakening of Papal power. This freeing of indigenous domestic aspirations from the previously commonly held notion of the existence of a voluntary and shared Christendom transformed social relationships in European countries and directly contributed to new modes of economic production, trade, exploration and scientific discovery.[16]

The freeing from a religious/political institution in Rome attempting to reconcile and advance essentially theological, epistemological and ontological considerations with a rapidly expanding European political system facilitated trade and exploration patterns which were far advanced by the time the ex-Augustinian monk Martin Luther nailed his 95 theses on the door of the church in Wittenberg in 1517. Henry VIII of England's break with the Church over the issue of divorce in 1534, and the defeat of the Spanish Armada in 1588 by Henry's daughter, Elizabeth I, brought to a close Catholic opportunities of overturning the new political and theological doctrines.

A new world was literally opening up across the seas. New peoples and new economic opportunities burst upon the European scene after Columbus made his discoveries beginning in 1492. These discoveries later solidified the massive changes taking place since the beginning of the Middle Ages.[17] Increased trade, greater wealth, the development of new and expanded social classes–all of these developments are traceable to the Reformation.[18] The institution of this modified set of political relationships, not only of king to pope, but between the aristocracy, merchants and other social classes (principally the peasants) altered the extent and degree of exploitative mechanisms applied to the physical environment. The Protestant Reformation ushered in a new set of social forces which began to undermine environmental constraints.[19]

It is not the purpose of this study to delineate all the various strands of historical, economic, social, theological and indeed psychological factors which contributed to the change in the relationship between human beings and their natural environment.[20] It is important to note, however, that the predominant philosophical and ethical position adopted by Western civilization that nature, the biological diversity and the physical and chemical processes that encompass it, is a threat to the human species and hence must be ruthlessly dominated as well as exploitated, pre-dates the revolution in religious thought which characterized 16th century Europe. As humans evolved from the natural environment, that environment was always a demanding and often cruel regulator and competitor for resources. For eons, existence on planet earth has been a continuing competition between environmental constraints and human needs. The human involvement in this complex biological system has been there since the very beginning of the species. But it has also thoroughly changed over time. In the pre-historic stages ante-dating the beginning of human civilization in the valleys of the Tigris and Euphrates rivers, the capacity to alter the environment, much less fundamentally transform it, did not exist. Even with the development of agriculture 10,000 years ago the human impact on the overall planetary environment was nil. Indeed, since current archaeological evidence has not found civilized human settlements older than 10,000 years, human impact on the overall planetary environment was impossible. Paleo-anthropologists tell us that hominid species existed roughly 3 million years ago,[21] but the age of the earth itself is several billion years old. Yet, in the space of a very few centuries, just seconds on an allegorical geological time clock, humankind has itself come to threaten life on its home planet.

Three Great Revolutions

There have been three great revolutions in social life, events so important that all subsequent human social existence has been organized around them. The first was the *Agricultural Revolution*. This was the classic transformation from bands of hunter gatherers to crop growers, planters and harvesters. One hundred centuries passed before the next great change, and then this occurred only some two odd centuries ago from our present age. An *Industrial Revolution* originating in England spread and rapidly advanced to create a global economic order. This world-wide production, distribution and trading system grew especially rapidly during the current century and particularly during the last 45 years. In the space of less than two generations, a global transformation of such momentous and far-reaching consequences has spread to every locale on the earth. The third transformation, the *Ecological Revolution*, is the one currently occurring. This is also a far-reaching change, but the challenge to the human species this time is to attempt to reconcile exploitative and short-term mores with concepts of sustainable development which are ecologically sound.[22] This present work will ask whether such efforts as being undertaken by nation-states, intergovernmental organizations (IGOs), international non-governmental organizations (INGOs) and other interested parties or groups regarding protection and improvement of the global ocean are likely to meet with success.

It must be noted that the second revolution, the industrial revolution, like the first, the agricultural transformation, was not simply a process of negative outcomes. Far from it. Human beings reaped huge benefits from each of these transformations. The industrial revolution, however, altered the physical environment much more than its predecessors; and it is the problems generated by this industrial paradigm that are such a challenge to the world community today.

There seems a great reluctance in many quarters to discuss questions related to the future role and function of science and technology in contemporary life. To be sure, a vast literature has developed in this field. Yet the question must be asked: Has this contemporary body of evidence significantly impacted upon the policy outcomes of most national governments, as they effect the ocean, in the West? The answer is no. Why this is so will be discussed in the following chapters; however, some reasons can be stated here. The reasons why governmental policy does not assign a predominant role in national decision-making processes are multifaceted. It is felt in some cases the issues are by their very nature so complex as to deny the utility and perhaps the possibility of a vigorous and thorough questioning of basic theoretical foundations. Exceptions may be made for those specifically trained and socialized into the formal structures of a scientific discipline; but most politicians at the national level, of course, are not scientists. And the obverse holds equally true, most scientists are not elected politicians. This is dangerous for a democratic society; not that scientists are not politicians, or that politicians are not scientists. What is dangerous is that avenues or mechanisms for transferring the knowledge discovered by intellectuals is so primitive or rudimentary given the fact that the nature of the threats to society are so basic.

Heilbronger and Kevles have noted the fallacy of this divorce between science and technology, on the one hand, and the intimate involvement these issues exert with the question of daily decision-making at the national level.

> It could be that neither the writers of textbooks nor their readers think they are knowledgeable enough about science and technology to deal with them... To understand the role of, say, agriculture in American life, teachers and students do not have to understand the intricacies of planting and harvesting. Nor do they have to master the details of modern science to understand its causes and consequences. What teachers need to know, and their students need to learn, is what might be called the "civics" of science and technology—how the two work as institutions, how they relate to society, and how they influence government actions and policies.[23]

This quotation raises several interesting issues. Here, political science can play a very useful role. This section will lay out some of the basic problems science and technology present to the political process, especially as they relate to industrialized market economies and the transformation those states are currently undergoing.

Earlier in this chapter the definition of the exclusionist world view of Dennis Pirages was presented. There is another outlook possible as well, one Pirages has labeled the *inclusionist perspective*.

> This ... focuses on humankind as one among many species competing for the earth's limited resources... The perspective ... stresses the pressing nature and sociopolitical impacts of global issues.[24]

This analysis on the impact of the extant industrial ideology and the interactions of it on industrial man and woman[25] are critical, and presage the environmental problems faced by humanity; particularly in the ocean.

What are the variables that go to make up what can be called the *Pirages Paradigm*? The exclusionist world view is materialist, in the tradition of the writers of early, modern and post-modern capitalism.[26] The dominant values of this paradigm are those of the unregulated or minimally regulated market; and its advocates propose limited governmental management or involvement with the economy.[27] Inclusionists do not see the challenges of the next century being met successfully this way. To quote again from Pirages:

> Inclusionists do not see an unregulated marketplace as an ideal way of meeting human needs, given the seriousness of environmental abuse, and they argue that economic activity must be managed carefully if destruction of nature is to be prevented. This nascent inclusionist world view is offered as

> a post-industrial competitor ... [to the current dominant thinking of industrial society].

The inclusionist theory gives high priority to what are described as "techno-ecological" variables, and the effects these have on social structures and values.[28] In the last decade of the 20th century the most important explanation for what is going on in the international system seems to be these "ecological and technological variables" that are manifesting themselves in resultant social, ethical and political challenges. In other words, an analyst employing the tools of contemporary political science would find greater heuristic value in those approaches which emphasized environmental and scientific and technological variables to more accurately understand the current international system than the more traditional approaches of balance of power, neo-realism, et al.[29]

The great value of the Inclusionist Perspective for international relations theory lies in its holistic approach. It incorporates structures not usually given wide discourse in the literature. One such example is in the similarities between human evolution and the evolution present in social systems.

> Just as [physical] bodies of contemporary human beings represent the products of hundreds of thousands of years of biological evolution, social structures and collectively held values are to a great extent the products of a parallel social evolutionary process. The human body has adapted to a changing physical environment through natural selection processes based on differentiated rates of reproduction. Those genotypes best adapted ... have survived... Social evolution is likewise a dynamic process by which values and forms of social organization adapt to changing technological and physical realities.[30]

This is a new way to view and study international politics. It took a very long time for the modern human species (homo sapiens sapiens) to come into existence. Part of the "social carryover" which concomitantly traveled with us on this journey were the associated learning mechanisms which humanity, over the eons, came to appreciate as efficacious or functional to a strategy of life and survival. Cooperation *and* conflict, community organization (especially in political matters) *and* the quest for a greater and more profound understanding of the world around it have all fueled human adaptation, growth and patterns of interaction. These are the components of the Ecological Revolution which are now getting underway.

The Transformation and Dissemination of Knowledge

Before discussing briefly the impact science, technology and communication exercise on contemporary international developments, it is useful to remember certain and often overlooked components to the formal quest for knowledge.

Knowledge is very complex, because it is essentially an epistemological quest. Knowledge is related to language in that human beings cannot exercise the fruits of their knowledge without it. Indeed, the entire investigatory process cannot proceed without acknowledging the somewhat tenuous connection between what are called "words" and what is "real meaning."[31] It has been noted, for example, that any useful or worthwhile attempt to address this question has to differentiate—admittedly very difficult—"between verbal meaning as the denotation of words, and material or 'real' meaning as the nature of things ... we wish to understand." Just exactly what meaning *is* lies in the connection between these two strands. Morehead Wright succinctly addresses the issue; without however, it must be added, entirely resolving it.

> We cannot formulate and express real meaning without the use of words or other symbols, so that language in large part structures our thinking about the nature of things. Yet the objects of thought are strongly resistant to our efforts to pin them down in language.[32]

It is a truism to note that in the 20th century science, technology and communication have transformed human life more quickly and more thoroughly than at any other time in history. In the chapters which follow, the transformation and dissemination of knowledge through the contemporary international system will be discussed. This is necessary because history itself, it seems, has accelerated at an ever increasing pace; especially since 1945 and the dawn of the age of space exploration in 1957. It would belabor the point to document the changes here. So, what can be said about these changes—in other words, what of some of their effects?

Progress in this field is a two-edged sword, as many observers have pointed out.[33] What has passed for "advancement" or "betterment" has concomitantly increased costs across vast areas of human life. This has been true planet-wide, and as will be discussed later, notably in the oceans.

There has certainly been a positive side to scientific discovery and its applications. "The positive side ... has been the industrial revolution and higher living standards for the relevant populations... The march of technology has yielded new capabilities [but it has also] created commons and collective goods problems."[34]

The "tragedy of the global commons" was first discussed by Garrett Hardin in a path-breaking article in 1968.[35] In it he postulated the interconnected components of economic, social and environmental factors by addressing the issue of non-sustainable exploitation of natural resources. Simply put, if an area of social worth (say an area like a village commons in 17th century England) involves several parties who use the piece of ground for economic purposes (such as grazing cattle or sheep), and are not to destroy the group (or collective) benefits brought about by the mutual sharing and support provided by the commons; then private, short-term, selfish interests must be controlled and regulated. Otherwise, the economic pressures become too great to bear and productivity on the land collapses. At its essence this regulative effort involves measures of behavior alteration, including legal enforcement mechanisms. Hardin states if the exploitation of a commons is allowed to go unchecked, then biomass collapse ensues. Individual as well as collective attitudes and values are involved. This analogy can be applied to a number of issue areas.

The term "social trap" is a useful heuristic device for sensitizing and revealing to analysts and policy makers the dilemmas confronting both modern Western society and in an even larger sense the extant Global Village—a unique creation of the telecommunications age.

The phenomenon of increasing dispersal of scientific and technological knowledge is having a destabilizing effect on national governments and on their interactions within the international system. Knowledge is not closely held or tightly confined within national borders. It is an amorphous entity, spreading its influence everywhere. Electronic telecommunication plays its large role, but so too do international congresses and meetings, publications in professional journals, and personal contacts and correspondence. At its essence, science is international. It is not territorial, national or even personal. The old adage holds true now more than ever: Science knows no boundaries.

The nation-state, however, is not similarly organized. There, of course, the guiding principles are those of territoriality and boundaries, sovereignty and national interest. This produces serious anomalies which will be discussed in the chapters to follow. Perhaps no where else are the effects so debilitating as those found in the field of environmental coordination and management questions. Chapter 1 will explore in detail the theoretical background underlying the current state system.

IGOs and MNCs

This preamble or set of preliminary observations on the contemporary state system has briefly examined three areas or issues antecedent to the current challenges confronting the global system. They are (1) the problem of development and progress; (2) scientific and technological knowledge as it relates to the international system; and (3) the impact of communication technologies on world politics. There is a fourth and essential concluding component, and that is the role of intergovernmental organizations (IGOs) with multinational corporations (MNCs).

There are many international organizations in the world today (both IGOs and INGOs), performing a wide variety of functions or tasks.[36] The extensive scope of activities will not be addressed here. This section, instead, looks briefly at one such interaction: the role of the United Nations and its subsidiary or affiliated agencies with the activities of the global corporation.

Multinational corporations exercise a tremendous influence within the international system. Many are organizations with tens of thousands of employees, and billions of dollars in sales and assets. Many MNCs have yearly total sales greater than the Gross Domestic Products of many nations (mostly, of course, in the Third World). With legal headquarters usually in specific countries, their operations extend world-wide.

Nonetheless, there are international negotiations which have sought to regulate MNC operations.[37] In addition, the United Nations secretariat has a Centre for Transnational Corporations.[38] (The term multinational and transnational are used here interchangeably.) The essential point, though, is the current status of the MNCs functioning as a global actor. Their importance has, by now in the 1990's, been clearly axiomatic for at least the last three decades.

The problem of pollution in Third World nations caused by these resident MNCs is a serious one. It is not the only source by far of pollution in Third World nations, however. Indigenous corporations, owned and operated by nationals, contribute to the problem as well. However, the role of the outside corporations has received widespread attention within the international relations field, and within the United Nations system itself.[39] There have been strong efforts, in years past, to develop a code of conduct for MNCs. H. Jeffrey Leonard has written that

> [T]he Commission on Transnational Corporations of the U.N. Economic and Social Council include environmental factors in its code of conduct for multinational corporations. The draft code calls on corporations to disclose to the authorities of the countries in which they operate all relevant information concerning features of their products or processes which may harm the environment.[40]

Multinational corporations have come to play important and decisive roles in the exploitation of deep sea-bed minerals and other potential resources of the ocean. This is more in potential than in current activity. Due to the failure of the Law of the

Sea Regime to enter into full existence as envisaged by the treaty, MNCs are not playing the role which was envisioned for them. Indeed, it was a disagreement over the proper role of the MNCs in the proposed International Sea-Bed Authority, as defined by the Government of the United States, that led it to reject membership in the proposed international organization.[41] This has proved a very serious obstacle in creating an effective regime.

This extended, but preliminary, discourse is offered as an introduction to a few particularly vexing and salient issues. It is meant to lay the groundwork for the analysis to follow.

Chapter 1 Notes

[1]David W. Chu and Ware Morehouse, "Third World Cooperation in Science and Technology for Development" in Volker Rittberger, ed., *Science and Technology in a Changing International Order, the United Nations Conference on Science and Technology for Development* (Boulder: Westview Press, 1982).

[2]See Michael Walzer, *Spheres of Justice, A Defense of Pluralism and Equality* (New York: Basic Books, Inc., 1983).

[3]For purposes of this book the West is defined as the industrial market democracies in Western Europe, North America and Oceania.

[4]The dynamism of civilization in East or West is not what is at issue here. The point to be made is a simple one. Namely, the interaction between the variables or components of Western civilization have produced a greater material base and scientific advancement than in the East.

[5]Prime examples are Khomeini's Iran, certain Persian Gulf states (including Saudi Arabia), Mao's China during most of his rule, Albania under Hoxa and Cambodia under Pol Pot and the Khmer Rouge.

[6]There are, of course, competing designs offered by the Marxist critique; as well as examples of certain Third World leaders who (for a period of time, at least) withdrew their nations from the international economic system (Tanzania under Nyerere). Other nations, such as Albania, have been non-participants in the international economic system since the end of the Second World War and the beginning of the Cold War by choice.

[7]The term "underdeveloped countries" is in many ways a misnomer. Dag Hammarskjold once remarked that "underdeveloped ... [is] a rather curious term ... if you realize that a few of them have civilizations dating back far beyond anything we know of... [T]hey have attained a maturity of the individual which makes you feel rather humble. They are underdeveloped only in the economic sense." Quoted in Brian Urquhart, *Hammarskjold* (New York: Alfred A. Knopf, 1972), pp. 375-376 (From Hammarskjold's unnumbered papers, 30 March 1954).

[8]See John Gerard Ruggie, "On the Problem of 'The Global Problematique:' What Roles for International Organizations?" *Alternatives* (May 1980) and *Our Common Future, The World Commission on Environment and Development* (Oxford and New York: Oxford University Press, 1987). For a somewhat dated, but still useful, laying out of the basic relationships in the North-South paradox see Richard W. Sterling, *Macropolitics, International Relations in a Global Society* (New York: Alfred H. Knopf, Inc., 1974); particularly Chapters 12 and 24. Also, see Lester R. Brown, *World Without Borders* (New York: Random House, 1972); Harold and Margaret Sprout, *Toward a Politics of the Planet Earth* (New York: Van Nostrand, 1971); and Gunnar Myrdal, *The Challenge of World Poverty* (New York: Pantheon, 1970).

[9]This says nothing of the problems posed by increasing population. See especially Lester R. Brown and Christopher Flavin, "The Earth's Vital Signs," in *State of the World 1988* (New York and London: W.W. Norton and Company, 1988); and Lester R. Brown, "Analyzing the Demographic Trap," in *State of the World 1987* (New York and London: W.W. Norton and Company, 1987).

[10]The seminal work in this field was done by Thomas Kuhn. See his *Structures of Scientific Revolutions*, 2nd ed. (Chicago: University of Chicago Press, 1970).

[11]Dennis Pirages, *Global Technopolitics, The International Politics of Technology and Resources* (Pacific Grove, CA: Brooks/Cole Publishing Company, 1989), p. 8.

[12]See Owen Chadwick, *The Reformation* (Middlesex, England: Penguin Books, Ltd., 1977), especially Chapter 1 "The Cry for Reformation" and Chapter 11 "Divided Christendom." Another excellent study is presented by Justo L. Gonzalez, *A History of Christian Thought, Volume III, From the Protestant Reformation to the Twentieth Century* (Nashville, TN: Abingdon Press, 1980), especially Chapter 2 "The End of an Era" and Chapter 12 "The Changing Philosophical Setting." Finally, see A. G. Dickens and John Tonkin, with Kenneth Powell, *The Reformation in Historical Thought* (Cambridge, MA: Harvard University Press, 1985), especially Chapter 1 "Views from Within: Early Historians and Controversialists." These authors (Dickens, Tonkin and Powell) perceptively note "Already in its early years, the Protestant Reformation was seen as something more than a religious dispute, and the sheer diversity of its effects has since been observed with ever-increasing coverage and precision," p. 1, op. cit. A few sentences later they add: "We shall not contrast the social sciences and theology as two mutually exclusive spheres, for after all religion exists within actual human societies, not within some spiritual stratosphere."

[13]See A.E. Goerner, *Peter and Caesar, Political Authority and the Catholic Church* (New York: Herder and Herder, 1965). Also, Joseph R. Strayer, *On the Medieval Origins of the Modern State* (Princeton: Princeton University Press, 1970). See Steven Ozment, *The Age of Reform, 1250-1550, An Intellectual and*

[26]See H. See, *Modern Capitalism: Its Origin and Evolution* (Philadelphia, PA: Ben Franklin, 1928).

[27]Dennis Pirages, *Global Technopolitics, The International Politics of Technology and Resources* (Pacific Grove, CA: Brooks/Cole Publishing Company, 1989), p. 8. This section is drawn from here as well.

[28]Dennis Pirages, *Global Technopolitics, The International Politics of Technology and Resources* (Pacific Grove, CA: Brooks/Cole Publishing Company, 1989), p. 9.

[29]Refer to John A. Vasquez, *Classics of International Relations* (Englewood Cliffs, NJ: Prentice-Hall, Inc., 1986); also James E. Dougherty and Robert L. Plaltzgraff, Jr., *Contending Theories of International Relations, A Comprehensive Survey,* 2nd ed. (New York: Harper and Row, 1981). Also see Richard W. Mansback and John A. Vasquez, *In Search of Theory, A New Paradigm for Global Politics* (New York: Columbia University Press, 1981).

[30]Dennis Pirages, *Global Technopolitics, The International Politics of Technology and Resources,* op. cit., p. 10. See also Marion Blute, "Sociocultural Evolution: An Untried Theory." *Behavioral Sciences* (January 1979) and Kenneth Boulding, *Ecodynamics: A New Theory of Societal Evolution* (Beverly Hills, CA: Sage, 1981).

[31]This section is drawn from Morehead Wright, "The Problem of Meaning in International Thought," in Michael Donelon, ed., *The Reason of States, A Study in International Political Theory* (London: George Allen and Unwin, 1978), p. 92.

[32]Ibid.

[33]This statement, axiomatic as it is, must be one of the oldest inherited bits of wisdom humankind has to offer. Its antiquity, however, in no way lessens its value.

[34]Dennis Pirages, *Global Technopolitics, The International Politics of Technology and Resources,* op. cit., pp. 132-133.

[35]See Garrett Hardin, "The Tragedy of the Commons," *Science* (3 December 1968). See also Garrett Hardin and John Baden, eds., *Managing the Commons* (San Francisco: W.H. Freeman, 1977), quoted in Pirages, 1989, op. cit., p. 170.

[36]Warren J. Feld and Robert S. Jordan with Leon Hurwitz, *International Organizations, A Comparative Approach*, 2nd ed. (New York: Praeger, 1988), p. 7. Feld and Jordan note that the number of IGOs may reach 450 by the year 2000 (378 in 1986 alone), and by the year 2000 it may be possible there will be over 10,800 INGOs (4,676 in 1986).

[37]Feld and Jordan, op. cit., pp. 269-274.

[38]See *United Nations Handbook 1989* published by the Ministry of External Relations and Trade, Wellington, New Zealand, 1989, p. 105.

[39]See particularly the following: John Tessitore and Susan Woolfson, eds., *Issues Before the 43rd General Assembly of the United Nations* (Lexington, MA:

Lexington Books, 1988), pp. 89-97; John Tessitore and Susan Woolfson, eds., *Issues Before the 44th General Assembly of the United Nations* (Lexington, MA: Lexington Books, 1990), pp. 96-100.

[40]This section is drawn from H. Jeffrey Leonard, *Pollution and the Struggle for the World Product, Multinational Corporations, Environment, and International Comparative Advantage* (New York: Cambridge University Press, 1988), pp. 54-82, especially pages 79-81.

[41]For an interesting reading of the Law of the Sea negotiations (from 1958 through 1982), see Clyde Sanger, *Ordering the Oceans, The Making of the Law of the Sea* (Toronto, Ontario and Buffalo, New York: University of Toronto Press, 1987), especially chapters 1-4.

2

Theoretical Aspects of the Current State System

Introduction

Planet Earth contains approximately 170 states. All formally claim independence and sovereignty. As such, they admit of no higher authority than themselves in matters of domestic or international concern to their respective governments. This chapter will outline the organization of the current state system. Theoretical concerns, drawing from writers of historical importance, will receive primary emphasis. Following this, the discussion will move to the contributions of contemporary international relations theory as it relates to an explanation of current *and* global politics. An evaluation of the latter position, within the field of international relations studies, will be offered. Throughout, the concern will be to provide an efficacious interpretation of the changes impacting the nation-state system.

Sovereignty

The theoretical foundations of the state system[1] did not simply spring into existence some specified day 3.5 centuries ago, in the middle of an era turbulent and scientifically rich with discovery. It has been noted, Martin Wight argued, "that the significance of the Congress of Westphalia [circa 1648] is seriously misjudged when it is taken to mark the 'beginnings' of the state-system rather than its 'coming of age' after some two centuries or so of improvised development."[2] The influence of history, its heuristic value for scholarship, in understanding the development of peoples and in the study of international relations, is axiomatic. One student of the field has observed

> [T]he scholar ... makes history the first and continual material of ... [his or her] science. History ... is not just an optional subject, something that historians do. It is not just a vital, passionate quality of the thought that

> makes the data of ... [the] science. History is the place ... [where] the search for patterns of regularity, order [and] principle ... [are sought].[3]

The utilization of an historical methodology for the study of international relations provides guidelines in the search for "coherence,"[4] coherence in a world radically diverse. Yet at the same time each and every act, whether by an individual or group (including nation-states), does not exist only in the realm of solitary isolation. Otherwise, no systematic study of this or any other issue would be possible. Discoveries of relationships or systems, embedded in the social fabric of the intercourse between nations, provides the dynamic element in the study of global politics.[5]

The elemental component to understanding the current state system lies in an examination of the term sovereignty. Unfortunately, scholars are not fully agreed upon just what this most important concept means. Alan James has perceptively noted the emotive nature of the word itself.[6] The resulting inexactitude has plagued the study of the state system since its modern inception in the mid-17th century.

James believes a fundamental reason for this lack of widespread understanding or consensus is due to an over-emphasis on the "internal nature"[7] or internal characteristics of the state. By this, he refers to sovereignty as a quality or essence that is possessed by the state.

> What sovereignty refers to at this level is the presence, within a governed political community, of supreme legal authority—so that such a community can be said to possess sovereignty, or to be sovereign, if it does not look beyond its own borders for the ultimate source of its own legitimacy.[8]

Another definition of sovereignty is offered below, but first it is necessary to briefly discuss some background to the development of the term in Western history (and thus contribute somewhat to the extant confusion within the field). That the idea of sovereignty was "developed" in the first few centuries of the Christian era,[9] and that the Roman Empire embraced this concept of the "internal nature" referred to above, was significant for later developments. The fact it was subsequently lost to future generations, until the early modern era during the Protestant Reformation, is also indicative of the changing nature and character of political organization. The relative harmony of views engendered by the Roman understanding did not last into the Middle Ages. Competition from an emerging and differing political tradition made itself felt in the form of a universal ideal of a shared Christian commonwealth. This concept of Christendom, present and operating within the existent political context and tradition, exerted a debilitating influence on the pre-Middle Ages conception of political authority, legitimacy and power. The political system of Pope and Holy Roman Emperor, along with kings in pre-modern nation-states, subverted and replaced the earlier understanding on the "internal nature" of sovereignty possessed by the Roman Empire until it made a reappearance in the 17th century.

As noted above, scholars do not fully agree on the meaning of sovereignty. Yet, there are some common points of accord. First, all the extant states on the planet today are "territorially based ... and at the international level ... [represent] it exclusively."[10] And, second, these states contain not just territory but people as well.

Efficaciousness and parsimony are essential components to definitional clarity. This approach is especially necessary when addressing a term such as the one under discussion here. The Permanent Court of Arbitration in 1928 defined the issue in the following manner: "Sovereignty in the relations between states signifies independence. Independence in regard to a portion of the globe is the right to exercise therein, to the exclusion of any other state, the functions of a state."[11] I would add that sovereignty once achieved is a matter of degree. No state on the planet possesses full or complete sovereignty; likewise, no state possesses full or complete independence from the influences or constraints of other states or from domestic factors. Geographical and demographic trends, and economic and technological variables, are just a few of the constraints that can operate deleteriously or favorably upon a state.

Sovereign Statehood reiterates sometimes overlooked or forgotten meanings to the term which are expedient to the issue of concern. An individual cannot physically go to a certain spot and literally see a sovereign state. One can see the *activities* of such a state, but one cannot see sovereignty per se. An essential difficulty in the study of sovereignty is thus explicitly exposed–that of the inexactness of the term itself. "[T]he absence of a physical referent has the consequence that one cannot tell those interested to go and have a look at the object in question..."[12] In the sense indicated here, indirect study of the object is thus the only avenue open.

Most states act on the international level to some degree, and the "gate-way" to this enlarged domestic governmental functioning is constitutional independence. "[T]he sovereignty on which their international activity is based amounts to constitutional separateness. A sovereign state may have all sorts of links ... but the one ... link, by definition, it cannot have is a constitutional one." This legal separateness is critical for James because it makes possible the practical involvement in international life with which he is concerned. Whether all states that have this possibility choose to exercise it (more accurately, whether they can in fact exercise it due to constraints of physical size or population or economics or other influencing factors) is another question. In *Sovereign Statehood* "sovereignty, in the sense here discussed, consists of being constitutionally apart..."[13] But this quality or condition is simply a prerequisite for admission, not a ticket for active or influential participation or even significant international presence or behavior. Although "constitutional independence" is the critical term used to define sovereignty[14] the definition is minimalist in the sense it denotes a state's basic legal document (whether it is joined together in one fundamental law as in the constitution of the United States or whether it exists by convention as is true within the United Kingdom does not matter to the fundamental thrust of the argument). It is thus clear that sovereign states, as the term is used today, cannot be a subsystem of larger constitutional classifications.[15]

While the interested student cannot see the activities of a sovereign state, the primacy of the territorial concept itself has received widespread attention within the field. The existence of territory, however small in size, is a litmus test for a state's admission and participation within the international system. (The role of international nongovernmental organizations, which do have a significant role to play within the international system, will be taken up in chapter 4.)

In the more traditional composition of the state system comprising territorially based sovereign states exceptions do exist; for instance, the recognition by some states of governments-in-exile. Examples here would include Polish and other East European states whose territories were occupied during the Second World War by German armies. The question here, though, turns more pointedly on which government is legitimate in the eyes of the international community, or large sections of it, and which is not. German occupation of Poland in 1939 was not recognized by Britain and other states as a legal act under international norms and laws, and a Polish government-in-exile was recognized and established in London. In more recent times, the case of the Palestine Liberation Organization (PLO) provides a relevant example. This association of terrorist groups does not possess a territory where it exercises the functions of sovereignty. Many member states of the United Nations, however, treat the chairman of the organization as a de facto head of state and have many of their public relations (an important caveat) with the organization on a level at par with those of established governments holding internationally recognized territory and borders. The United Nations have recognized the PLO as the "sole legitimate representative of the Palestinian people" which, in everyday practice, closely amounts to an ad hoc granting of sovereignty *minus terra.*[16] It is not particularly the purpose here to analyze the relative merits or lack thereof of the PLO. The point is simply to identify a relevant example where the lack of territory does not prevent a political organization and people from achieving international recognition. Many Arab and European states (in both the East and the West) during the 1970s and 1980s fall into this category of conducting extensive diplomatic intercourse with a group claiming statehood but not possessing land.

Ingrid Detter DeLupis has noted "[t]he territory of a state is the framework within which the state exercises its competence: the territory could even be said to furnish the very title of that competence. To have a territory is, in fact, one of the conditions of statehood..."[17] Statehood in this modern understanding of the definition, then, requires territory. The size of the territory can relate to the degree of sovereignty enjoyed by a nation-state (consider the examples of the United States, the Soviet Union, the People's Republic of China, or even on a smaller scale, the United Kingdom, France or the Federal Republic of Germany), but this is only a very rough and inexact indicator and more often it is not. Again, consider the examples of Chad or Mali, on the one hand fairly large states in terms of territory, and Belgium or the Netherlands, which are much smaller in size, yet hold a much greater level of influence within the international system. We can say these Benelux countries possess "more" sovereignty than the geographically larger African nations. Many things separate these two latter groups of nations, but one of the most obvious is size of gross national product (GNP). Other indicators besides size of territory could be offered for the first group of European nations and China, but the problem would still be just as difficult to measure. Many examples similar to this could be offered.[18] Suffice to state, while territory is a prerequisite to formal statehood, size of territory or even its existence is not a *sine qua non* for influence within the international system.

F. H. Hinsley has perspicaciously written on the question of sovereignty. It is a very unique kind of reflection. It is one which extends its influence through international politics in an anterior manner, as a sort of memory-trace. To carry this analogy somewhat further; it is lineal, although not direct, and it is not tenuous. Hinsley observed

> [Individuals] ... do not wield or submit to sovereignty. They wield or submit to authority or power. Authority and power are facts as old and ubiquitous as society itself; but they have not everywhere and at all times enjoyed the support or suffered the restraints which sovereignty, a theory or assumption about political power, seeks to construct for them... [S]overeignty is not a fact. It is a concept which [individuals] in certain circumstances have applied–a quality they have attributed or a claim they have counterposed–to the political power which they or other[s] ... were exercising.[19]

The expansion of these twin concepts of sovereignty and territoriality during the last three and a half centuries, eventually covering all land surfaces of the planet, contrasts strikingly with the lack of a similar historical development in the Global Ocean.[20] Compared with the previous land-based phenomenons of human discovery, settlement and consequent social transformation which were carried out through a long and at times disjointed process of historical development; a concomitant growth in the scope or breadth of law (whether domestic or international), and joined with the required apparatuses of regulatory and judicial structures operated by self-interested states, did not occur with respect to the realm of the ocean.

To be sure, and this needs to be clearly stressed, there has been an increase in the spreading-out, so to speak, of the type of land-based derivative authority referred to above, but only if the question is viewed through a very primitive quantitative prism—a perspective obviously devoid, to be sure, of interpretative nuance. It cannot be said sovereignty and law exist *only* on land. Such a statement is absurd—one needs only to consider the three-mile territorial limit in the history of international law or the required procedures associated with ship wrecks and salvage operations to demonstrate the preposterous elements of such a declaration. Simply put, therefore, "law" has of course operated in the ocean. It does so, however, in a way that is very dissimilar to that of the terrestrial areas on the planet–it is much less important as a concept and operative tool, and it is much weaker in terms of power, effect or scope to order and govern human society. This point will be explored further in the chapters to follow.

In addition to the expansion of these twin concepts of sovereignty and territoriality, three models can be heuristically employed to examine the issue of "control" which states have historically exercised over and within the Global Ocean.[21] The models laid out below are somewhat overlapping, although they primarily delineate three separate periods of state jurisdiction over the single largest physical component of Planet Earth. The three models are:

1. Freedom of the High Seas
2. Common Heritage of Humankind
3. Territorialization of the High Seas

The first model was the orthodox system of ocean control, jurisdiction and governance which existed for roughly five hundred years beginning with the age of the Portuguese and Spanish discoveries in the late 15th century and ending, exhaustively, in the latter years of the 20th century. This Freedom of the High Seas owes most of its strength to the writings of the 17th century Dutch legal philosopher Hugo Grotius.

> Grotius took as his basic premise that the oceans of the world were limitless in their bounty, incapable of being exhausted or destroyed by any human activity then imaginable. He noted further that the oceans formed the essential highway out of Europe to the world's frontier; therefore, he built his law of the sea upon a classic laissez-faire principle: that the oceans should be open to all capable of access to them on a mutually reciprocal basis.[22]

This interpretation held sway essentially until it was first seriously challenged in the 1960s by Arvid Pardo, the Ambassador of Malta to the United Nations. His now well known and influential call in November 1967 for the international community to declare the global ocean the "common heritage of ... [humankind]" ushered in the second model, this time with a proposed authority. Pardo's statement called for the creation of a substantive international regime to govern the seabed and ocean floor.

> Pardo spoke for an hour on the "known resources" of the continental shelf and ocean floor, ranging from archaeological and other sunken treasures in shallow waters to the wealth on the seabed and beneath. He also touched on the advances in oceanographic research...
>
> Pardo ... [called] for "an effective international regime over the seabed and the ocean floor beyond a clearly defined national jurisdiction," a regime that would not balk at facing the military aspects: "It is the only alternative by which we can hope to avoid the escalating tensions that will be inevitable if the present situation is allowed to continue. It is the only alternative by which we can hope to escape the imminent hazards of a permanent impairment of the marine environment. It is, finally, the only alternative that gives assurance that the immense resources on and under the ocean floor will be exploited with harm to none and benefit to all.[23]

Concomitant with this speech, and developing much more strongly in the following decade, the Third World became especially concerned at the growing inequality between economic development in their own states and the standards of living in the Western industrialized nations. Consequent and heightened pressures to industrialize, usually via the Western model, placed severe additional strain on an already overburdened planetary ecosystem. The Global Ocean could not be exempt.[24]

Pardo's vision was never fully realized. In large part because of the opposition of the Reagan Administration in the United States, the provisions of the UNCLOS (United Nations Conference on the Law of the Sea) treaty to operationalize the concerns underlying management of the resources of the ocean has never become a living reality. By the end of the first few months of the new administration (in March 1981) the head of the American delegation to the conference was dismissed and a wide-ranging review was held of the proposed components to the treaty. Leigh Ratiner has described the reasons for those opposed to UNCLOS.

> In essence, this point of view held that the treaty was flawed because it created adverse precedents for other negotiations on economic issues between developed and developing nations—the North-South dialogue—

> subjugated American industry to an international regulatory and management system, and was incompatible with President Reagan's apparent desire to return the United States to a period of power and influence in world affairs in which its policies would simply be enunciated rather than sold to others through a process of diplomacy and negotiation.[25]

The immediate background surrounding the emergence of a Law of the Sea concept can be traced to the first UNCLOS in 1958, and preceded Pardo's call for a transformational and philosophical change from world governments in how planetary resources were viewed.[26] Had his vision succeeded the world would have witnessed the creation of a viable and vigorous semi-autonomous international organization to govern the planet's ocean (particularly if the mechanisms for taxation were implemented). Such an International Seabed Authority (to be known as "the Enterprise"), which already has headquarters in Kingston, Jamaica, would have become a global resource management agency with significant powers of regulation.[27]

The third image of this heuristic design combined and expanded upon trends already in place for several decades previously. Because obsolescence had in many ways overtaken the first model, and unrealized hopes have handicapped full implementation of the second, many states moved quickly by the 1970s to adopt an expansion of the historical concept of sovereignty which previously had resided almost completely within land-based states and three miles outward from the coast as a maximum limit of a state's authority.[28]

The three mile territorial limit came under serious assault following the Second World War, beginning with the decision of the United States government in September 1945 to extend American sovereignty over the underwater resources of the outer continental shelf.[29] This eventually led other states, in turn, to declare a new standard of 12 miles as the jurisdiction of territorial waters and 200 miles as the jurisdiction for the newly created Exclusive Economic Zones (EEZs) which coastal states now began to implement.[30] It follows, therefore, that as coastal states have extended their sovereign reach the theoretical possibilities for enlarging the means for enforcing individual state laws and regulations over other parties has grown.

Contemporary International Relations Theory

Contributions by scholars in the field hold to a variety of interpretations on the question of what is happening within the contemporary international system. And certainly the more important question is: What *is* the international system? Systematic answers will be offered below, although they will not be definitive.

Approaches and methodologies differ, and this includes differences over which means provide the most powerful explanatory tools for judgment and interpretation of the international system. Consensus, much less unanimity, is not present. What can be asked of the discipline of international relations, and what has been asked for the past six decades, are a very basic set of questions. What are the most useful and heuristic propositions which can provide the interested student the answers, or at least partial ones, to the fundamental issues any science seeks to address: A) what is it we are actually observing, and B) what are the relationships and interactions between the components of the system?

Hedley Bull and Adam Watson have noted in the introduction to their excellent study on international society the essential importance of taking an "historical perspective" to the study of international relations.

> We certainly hold that our subject can be understood only in historical perspective, and that without an awareness of the past that generated it, the universal international society of the present can have no meaning.[31]

Contemporary international relations theory has patently drawn from the history of, what is commonly called, the international system itself. This term—international system—is more accurately described as international society by Bull.[32] What has made this "international society" a reality, if indeed one is present? It should be noted, not all international relations specialists have agreed that an international society is actually present.

Peter Gellman has written on the critical differences separating those within the discipline who hold to the existence of an international society. Some scholars have held that such an international society is not present, but rather what characterizes international relations is the reality of a "system of states." Most prominently among this camp is Hans Morgenthau, described by Gellman as believing a "system of states" is present and "bound together by fate."[33] This is very different from Bull's belief that the common "aims" and aspirations of states make for a "commonness" in goals and aspirations, hence a "society." As Gellman notes, "A political theory of international relations cannot accept a definition which refers to matters like protection from physical violence and respect for property, but not the question of purpose or the right to secure one's liberty. Purpose and not the formal inclusion or exclusion of states is for political realism a crucial consideration..."[34] In other words, objective standards or indices of power and its evaluation, as can best be defined by scholars, are the operative variables.

The Analyses of K. J. Holsti and Hedley Bull

Contemporary international society has been shaped by a series of vast historical changes. K. J. Holsti and Hedley Bull have analyzed these forces which have profoundly impacted on international relations theory.[35] Since Bull has identified four agents of this change, his contributions to an efficacious understanding of international society require examination, as does the work of Holsti. This section will conclude with a discussion of their common points.

> First, there has been the psychological or spiritual awakening of Asian and African, Caribbean and Pacific peoples, beginning among small groups of the Western educated, later affecting masses of people, that led them to perceive the old order no longer as a fact of nature, but as something that could be changed...

Here, Bull is speaking of the history of the colonialism (roughly from 1800 to 1955) and the effect the European Age of Exploration (also, roughly, from 1500 to 1800) exerted upon subsequent international developments. These have included forceful verbal advocacy for international political issues of Third World concern and interest. Admittedly, given current indices of power, both individual nations and to a large extent in multilateral forums as well, Third World nations have been particularly unsuccessful in achieving their goals of economic development.[36] However, the point raised by Bull relates to the realization by these various peoples that (1) their economic conditions are not acceptable, and (2) means are available to them to seek changes in patterns of trade and investment that have heretofore been the norm.

Bull notes further three fundamental alterations over the previous developing international society.

> A second factor has been a weakening of the will on the part of the Western powers to maintain their position of dominance, or at least to *accept the costs necessary to do so*.[37]

This observation is particularly obvious as any examination of post-Second World War international affairs will demonstrate. Beginning with the British decision to grant independence to India in 1947, to the French decision to withdraw from Vietnam after the debacle of Dien Bien Phu (1954), through the prescient "Winds of Change" speech of Prime Minister Macmillan in 1957, the main Western colonial powers pursued an inexorable policy of colonial withdrawal (often unwillingly, but not always, and eventually succumbing to the inevitable). The transformations of their world empires ensued.[38] American defeat in Vietnam and the defeat of the Soviet Union in Afghanistan are more recent and spectacular examples of European powers failing to impose their will on ostensibly backward nations.

> A third factor making for the demise of the old order was the impact of the Bolshevik revolution and the rise of the Soviet Union as a major power.[39]

This destabilizing impact of Soviet behavior on world politics has been extensive. The inversion of Western values by the Marxist-Leninist legacy significantly contributed to a breakdown of the traditional political order after 1917.[40] Aid to revolutionary movements has had a significant impact on transforming the international order. It is not the point here to adjudicate between the actions of individual sovereign states. It is clear enough condemnation can be shared by all sides.[41]

> A fourth factor assisting the efforts of non-Western states and peoples to transform the system has been the existence of a more general equilibrium of power, to which the rise of the Soviet Union contributed, but of which it was only part, that has operated to the benefit of those challenging the old order.

Interdependencies of national economies to global trade and production patterns have materially shaped and influenced this changing model of power.[42] The example of the oil crises during the 1970s is a pertinent indicator.

Bull's research provides strong evidence for the change within international politics and a replacement of one order by another. K. J. Holsti has also examined "distinct international systems" and their impact on contemporary international politics.[43] The years 1789 to 1939 and 1945 to the present (relevant, perhaps only up to 1985 with the ascension of Gorbachev to power) are, for Holsti, representative of separate international systems. The previous and current century have witnessed fundamental and basic changes "in the structures and process of international politics." Two distinct periods are identified.

> The first was the rise of nationalism—development of strong emotional attachments to the central state (adding to the traditional loyalties to provinces or towns) and involvement of the average citizen or subject in ... [his or her] governments' political life...

For Holsti, the development of nationalism was achieved through "ethnic or linguistic group[ings]," with the appeal to a heretofore not present ultimate and legitimizing factor for political organization—the nation-state.

Nationality is usually determined by place of birth. This accident of physical location generally makes one what one is–an American citizen, a French citizen, a British subject, a Zambian citizen or any one of numerous other existing nationalities.[44] It has, since the French Revolution in 1789 and the advent of Napoleon in 1799, become the most consistently powerful force within international politics.

> A second consequence of nationalism was that mass public involvement could be used by governments as a military and diplomatic bargaining capability. By invoking the theme of 'sacred national honor' governments could mobilize their populations to support their diplomacy or wars, while previously dynasts had found it difficult to generate popular enthusiasm for objectives not rooted deeply in social and economic aspirations.

This radical expansion of the scope of political discourse within nation-states and international society itself, due to the co-opting of masses of citizens, has led to an increase in the level of complexity required to manage the system. The number of actors rose correspondingly not only within the international system, but also within the various domestic systems as well. Holsti notes a critical shift within the European political arena with the introduction of the *levee en masse*—"a system of nationwide conscription designed to build a citizens' army." Prior to the French Revolution the tradition of "dynastic diplomacy" included professional soldiers recruited, trained and paid for by the king. The "symbolic break" with this system of the Bourbon monarchs occurred in 1791 when the French revolutionary leaders created a mass-based armed forces. "This army was sent abroad to 'liberate' Europe from dynasticism" such as had existed in France before the revolution. The army "... conquer[ed] Belgian, German and Italian territor[ies] for the greater glory of the French *nation*, not for the prestige of the French king." This was a crucial development. Such a subtle, yet profound shift would presage much of European and world history in the following two centuries by foreshadowing the rising power of nationalism as a key to state behavior.

The role of science and technology in the operation of war in nineteenth century Europe, combined with industrial prowess and development, was the second great transformation for international politics. This is an important shared component with the work of Pirages. He focuses, however, more attention on the implications and rule-making capabilities for international politics of science and technology variables (see Chapter 1). The use of these technologies, according to Holsti, added further to the "... symbolic break from the eighteenth-century tradition of dynastic diplomacy." Such complexity in warfare was highlighted on the European continent by the Prussian General Staff's utilization of transport and railroad scheduling during the Franco-Prussian War of 1870.

Holsti's work on the history of the European state system is an important contribution to understanding contemporary international politics and the creation of an international society. It is enduring in its contribution because of its utilization and identification of broad societal, political and historical realities that exert continued relevance to an understanding of contemporary international events.

Ideological Principles and Political Doctrines

A third transformation in the nineteenth century of the European state system "with major consequences on the structure and processes" of the European and, from that, the world political system "... was the rise of ideological principles and political doctrines as a major motive or guide to foreign policy behavior."[45] It is difficult to underestimate the break represented by the introduction and primacy of such an ideological component. It is, unfortunately, beyond the scope of this book to assess the impact of ideology, particularly ideologies of nationalism, on the state system and the global ocean other than to make a few preliminary observations.

Ideology is a formalized belief system. Myron Aronoff has done seminal work on the questions associated with culture and political change and has written of ideology as a specific subset of a political culture.[46] Aronoff sees great value in defining culture "as systems of shared symbols and meanings."[47] Drawing from Clifford Geertz and the work of Max Weber, Aronoff believes "... that ... [the human being] is an animal suspended in webs of significance [that the individual has spun] ... I take culture to be those webs, and the analysis of it to be therefore not an experimental science in search of law but an interpretative one in search of meaning."[48]

For Aronoff ideology is not static. It is changing, adaptable, preeminently culture-specific, yet bound by regularized patterns created from human interaction.

> In reality symbolic universes and ideology—which I view as representing a more specifically focused, conscious, and rationalized form of the political culture as it is manifest in the meanings to which political actors ascribe to explain and justify their behavior—contain ambiguities and contradictions, which allow for various interpretations... Various technological innovations, changes in the environment, socioeconomic and political contacts and changes can undermine accepted definitions of reality... The development of alternative definitions of reality can call into question the taken-for-grantedness of socially shared cultural meanings.[49]

Ideology can be rational. Unfortunately, however, history has often proven otherwise; but the possibility does exists for the belief system of the single individual, at least, to be as logically consistent as humans are capable of achieving. The issue is much more complex, however, when extrapolating from the rare person who systematically and rationally self-reflects on political values to larger groups and the sovereign state.

Those political ideologies present today, at least in the West, and in whatever form—whether that be industrial market democracies or socialist planned economies, and all the variations contained therein—have much to offer as a *disincentive* to the rational process of attempting to create formal belief systems that are pragmatic, workable in the sense of long term survival and viewed by those under its authority as legitimate. Perhaps human emotion and irrationality is the fundamental stumbling block to an intelligent and reasoned system of political beliefs. If so, the search will always be incomplete and the degree to which the search is incomplete may presage the ultimate success or failure of the proposed ideology.

Holsti has viewed the current international system, whose genesis was in the European political order, as differing in five significant ways from its predecessor. It is different because of

1. the rise in the number of states
2. great capacity for destruction
3. vulnerability of states to destruction and subversion
4. the rise of non-governmental and multinational political organizations–national liberation movements, multinational corporations, international interest groups and political parties ... and
5. the predominant position of influence that has been achieved by three essential non-European states, the Soviet Union, Communist China and the United States.[50]

Taking point five, first, several modifications must now be made to Holsti's analysis. Events within the Soviet Union and Eastern Europe during the winter of 1989 and 1990 have forever changed the post-war world. German unity, under terms favorable to the West, is an accomplished fact. Heretofore Communist states in Eastern Europe have either thrown the communist party out of power and are in the fundamental process of restructuring political life, or have held free elections which have resulted in the party losing its historic monopoly of power.[51] Either way, the basis of past influence for over forty years by the Soviet Union in this part of Europe has been overturned—and without war. But the appropriate political and economic system—and the values that would underlie it—has not yet emerged. This very uncertainty could be a factor for serious instability in the future.

Holsti's analysis, however, is still valid on the basis of broad historical and political trends put in place at the end of the Second World War. In this, the "position of influence" wielded by the People's Republic of China in world affairs still holds. Here too, though, great changes are restructuring international society.[52] The position of the United States has not remained static either, but more continuity has been present—in regard to international structures put in place after 1945—in its relationship with the European continent than that of the Soviet Union.[53]

The validity of point one is obvious. The number of nation-states has risen dramatically in the 20th century, as opposed to those existing in the century before. States possess in their armed forces and weaponry (particularly the thermonuclear weapons) the potential for catastrophic destruction. Yet, at the same time security has not increased and, as Holsti observes in point three, states in this century have become more vulnerable to "destruction and subversion." Finally, Feld and Jordan have investigated the growth, as Holsti describes them, of "non-governmental and multinational political organizations, national liberation movements, multinational corporations and international interest groups and political parties" and found their influence to be increasing.[54]

The Bases of the International System

The changed bases of the international system since 1945 has required analysts to try many approaches to understanding the reality of international politics. It is certainly not the intent here to state international relations were actually *understood* prior to this time. No age has ever possessed such a monopoly of wisdom in its own lifetime. Rather, it is the intent here to draw out the differences separating the post-war international system from the inter-war years of 1919 to 1939.[55]

One such approach utilized by Ernst Haas has been an examination of the International Labor Organization making use of functionalist theory, as developed by David Mitrany and others.[56] Haas aptly describes the issue at hand.

> Functionalists, in the specific sense of the term, are interested in identifying those aspects of human needs and desires that exist and clamor for attention outside the realm of the political. They believe in the possibility of specifying technical and "non-controversial" aspects of governmental conduct, and of weaving an everspreading web of international institutional relationships on the basis of meeting such needs. They would concentrate on commonly experienced needs initially, expecting the circle of non-controversial to expand at the expense of the political, as practical cooperation became coterminous with the totality of interstate relations. At that point a true world community will ... [arise].[57]

Functionalism can be criticized on many grounds. The most significant is in its down-playing and indeed denial of politics itself. The unintended, but unavoidable, consequence of this approach results in two ironies: (1) the theory albeit correctly sees that not all human needs are political or reside within the confines of political discourse; yet this accurate assessment moves to and concludes (2) that because all human needs and desires are not political that questions of management concerning even "non-controversial" aspects will not themselves generate controversy. When individuals disagree over proposed courses of action politics inevitably surfaces.

Haas concludes his extensive study with some clear lessons. "Neither Functionalism nor functional analysis can bring international order out of the chaos of national confrontation... But functional analysis can tell us in which direction the faint ripples of common concerns are likely to spread."[58] Functionalism as a theory, then, is like other approaches to the study of international politics: it is not complete in itself, but contributes to the increase of knowledge. In building upon its own contributions and pointing to other directions where further research can be undertaken, it is a fruitful endeavor.

Inis Claude has devoted considerable attention to the question of global politics—does it exist; if so, how and under what circumstances? If it does not, why not and might global politics exist in the future, however indeterminate?[59]

For Claude there is a critical difference between international *organizations* and international *organization*.[60] The former can be said to characterize large sections of international politics—i.e., there are many international organizations and many different kinds of nations which belong to them. Large global bodies such as the United Nations exist, and perform many political and ostensibly non-political activities or functions. Obvious political activities would include peace-keeping operations, and the airing of disputes within the Security Council. Ostensible non-political activities include the education and health care needs of Palestinian refugees under the aegis of the United Nations Relief and Works Administration for Palestine which reports to the United Nations Economic and Social Council. Other areas of similar involvement may include the United Nations High Commissioner for Refugees and humanitarian work in Pakistan for Afghani refugees, or in Africa attending to the needs of the many casualties that have resulted from numerable civil wars and traditional or human-made disasters. Claude perspicaciously notes that "[p]articular organizations may be nothing more than playthings of power politics and handmaidens of national ambition." At times, during the history of the United Nations organization, this definition has come very close to describing the current state of affairs. The decision of the Soviet Union to invade a member state of the Warsaw Treaty Organization in 1968 (Czechoslovakia) certainly falls under Claude's definition, as well might the action of the United States with invading Panama in December 1989 ostensibly declaring its intention, inter alia, to confirm in office a freely elected and democratic government which was prevented from assuming power by a military dictatorship. (Panama is a member state of the Organization of American States, as is the United States, and both nations are signatories to the original treaty which declares as inadmissible under international law the interference into the domestic affairs of one state by another.[61]) Further questions as to United States' disrespect for its treaty obligations under the United Nations Charter are raised by its decision to mine Nicaraguan harbors during the Reagan Administration. The International Court of Justice declared this action illegal and the United States promptly announced its intention to ignore the ruling.

International organization, however, represents a very different condition. It is "an historical process ... [It] represents a secular trend toward the systematic development of an enterprising quest for political means of making the world safer for human habitation." Claude makes a point with this analysis that is heuristically powerful. The point or lesson is this: The movement towards global politics on Planet Earth assumes the force of an inevitability. It is a natural or inevitable progression proceeding from the very beginnings of human social organization. This process enters back into the oldest eras of the collective social memory and traces itself up to our current times. The necessity of cooperation and more formal and hierarchical patterns of government (with a multitude of political forms—Kingships, Oriental despotisms, Syncretic Empires, Religious-cum-Political Commonwealths, Military Dictatorships and/or Technocratic-Bureaucratic Dictatorships; or a whole host of other variations produced by the human species and too numerous to name or detail here) has been present in human history.

As the numerous examples immediately mentioned above indicate, the "how" of political organization has varied tremendously. The point made by Claude, and to which I fully agree, is that despite all the twists and turns of human political life it is possible to trace this process of organizing upward from the single individual eventually reaching what we call today the international system. It began thousands of years before the modern nation-state entered into existence. Despite the massive changes in the history of social structures, international organization has represented a "trial and error" approach to the solution of political, hence human, problems. Claude notes this process "... may fail." It is not foreordained to succeed, and the identification of this hypothesis does not necessarily entail that the global organization spoken of will inevitably deepen or become ever more imbedded in the social fabric. Claude has identified a process and slightly nuanced it so that he cannot be accused of prediction without the necessary substantiating evidence. He qualifies the description by noting, "[b]ut if it maintains the momentum that it has built up in the twentieth century, it may yet effect a transformation of human relationships on this planet which will at some indeterminate point justify the assertion that the world has come to be governed..." Will this inherently be a good thing? This is a complex question to pose. It depends on how the process occurs and what it finally winds up becoming, if indeed it does "become" at all. As in many things, the ultimate answer will be given by time.

Before discussing critiques to the balance of power theory of international relations it is first necessary to discuss a representative sample of its proponents; those who hold it to be the most realistic and accurate approach. Under this rubric there are scholars who, while holding to the basic tenets of the theory, introduce modifications. The balance of power or realist approach holds wide sway within the discipline of international relations. Exponents include Hans Morgenthau, other scholars such as Peter Gellman, and latter day reformers including Robert Keohane and Joseph Nye, along with Michael Mastanduno, David Lake, and G. John Ikenberry. Morgenthau defines political realism in terms of six basic principles. They are:

1. Political realism believes that politics, like society in general, is governed by objective laws that have their roots in human nature...
2. The main signpost that helps political realism to find its way through the landscape of international politics is the concept of interest defined in terms of power...
3. Realism does not endow its key concept of interest defined as power with a meaning that is fixed once and for all. The idea of interest is indeed of the essence of politics and is unaffected by the circumstances of time and place... Yet the kind of interest determining political action in a particular period of history depends upon the political and cultural context within which foreign policy is formulated.
4. Political realism is aware of the moral significance of political action... Realism maintains that universal moral principles cannot be applied to the actions of states in their abstract universal formulation, but that they must be filtered through the concrete circumstances of time and place.
5. Political realism refuses to identify the moral aspirations of a particular nation with the moral laws that govern the universe... To know that nations are subject to the moral law is one thing, while to pretend to know with certainty what is good and evil in the relations among nations is quite another. There is a world of difference between the belief that all nations stand under the judgment of God ... and the blasphemous conviction that God is always on one's side...
6. The difference, then, between political realism and other schools of thought is real and it is profound... Intellectually, the political realist maintains the autonomy of the political sphere, as ... [other disciplines] maintain theirs.[62]

Morganthau's work was a seminal contribution to a rigorous understanding of the relations among states in a post-war world. The experience of the Second World War, the most violent conflict ever in human history, was profoundly disappointing to those who had labored and fought after 1918 for international mechanisms and controls to regulate inter-state conflict. The horrors of the First World War were plain and obvious to all men and women of good will. Morgenthau believed the political theory which dominated the discipline after this catastrophic event—political idealism—was a grave error and led directly to a new outbreak of fighting on the European continent a mere 21 years later. The results of that breakdown of international order between 1939 and 1945 unleashed a scale of death and destruction hitherto unseen. The way in which the war ended (it is, however, really more accurate to say it did not formally end because the victorious allies could not agree on the full terms of peace) led to the creation of a new period of world tension known as the Cold War. Naive misunderstandings of "objective laws that have their roots in human nature" (see point one above) helped create the theory of "political idealism" which Morgenthau devoted much of his scholarly career to refuting.

Peter Gellman has described this dominant theory of the inter-war years in this way:

> Political idealism, as it [first] emerged in The Enlightenment, promised to replace the quest for knowledge with knowledge proper. It claimed to have superseded Socrates and his apology for knowing of his ignorance with a system of knowing which made ignorance, in Morgenthau's description, 'a mere quantitative shortcoming' which might be 'overcome in due course of time.'[63]

The capture by political idealism of the mainstream in political debate during the 1918-1939 interregnum distorted international political life by denying the central importance of a theory of politics to understand the system of relations among states. This Morgenthau could not accept.

Morgenthau's writings on political realism have assumed an existence "larger than life" within the discipline and many things are ascribed to his views which he himself did not hold. This is unfortunate. A prime example is the belief, held by many who are only vaguely or superficially acquainted with his writings, that since interest is defined in terms of power this will always and everywhere mean military power exclusively. This is quantum leap in logic, unjustifiable; and yet an all too common plague on thoughtful and serious writers like Morgenthau who give careful consideration to their choice of words. (The term military power is here used to be the armed forces of a nation, supported by the domestic economic infrastructure.) Morgenthau was never so limited in his understanding of a concept this important. Gellman has observed

> Morgenthau did believe that as long as the world was organized along the lines of state sovereignty, the state would be 'the last word...' But the last word is a contingent word, and allegations of state-centrism notwithstanding, Morgenthau also argued that the notion of national interest was as transitory as the sovereign state that made use of it. 'The national state itself is obviously a product of history and as such is destined to yield in time to different modes of political organization.'[64]

The existence of thermonuclear weapons was, for Raymond Aron, the fundamental question of our time. As he wrote in the preface to the American edition of his work *Peace and War, A Theory of International Relations* "... China[,] the Soviet Union ... [and] the United States ... must seek an answer to the question that dominates the diplomacy of our epoch: how to use the thermonuclear weapons diplomatically so that they will never have to be used militarily?"[65] Aron developed a methodology that operates on four levels:

1. the schematic arrangement of concepts and systems
2. the general causes of events
3. the development of the sport or of a particular game [Aron uses the term "sport" or "game" as an analogy to denote regularized rules for political actors.] and
4. pragmatic or ethical judgments.[66]

This discussion will not go into an explanation of the details of Aron's system. However, certain points are in order.

Diplomacy is a vital concept for understanding the dynamic of the global system. This process is "cooperative and competitive," and the two major actors in the "practice of foreign politics" are the "diplomat and the strategist."[67] Each exists in a dual relationship that provides the driving force or engine for the reconciliation (or non-reconciliation) of differing national ambitions, goals, policies, aims and objectives. Actual fighting/combat or the maneuvering and perturbations of states below the threshold of armed conflict is, for Aron, one of the great paradoxes of modern international life. Unable to avoid the consequences of a state system founded on territoriality and national interest, the advent of scientific and technological variables post-1945 have introduced a fundamental threat to the continued operation of the entire system. The irony is how to avoid ecological—and especially marine—destruction when the means of destruction have increased so dramatically and exponentially (such "means of destruction," are not limited, as I have stated, to strictly military considerations alone).

It is not the intention of this chapter to pronounce on the work of the scholars here used to illustrate a certain point or to describe a broad historical development that relates to the contributions of the international relations literature. Personal evaluation is perhaps unavoidable in social science research, yet this does not mean opinion should be vaunted as fact or knowledge. Besides, and more to the point, evaluating the ultimate merits and misconceptions or emissions of any of the scholars used here would be a very different intent from the one envisaged. This chapter is intended as a selected overview on the extant contributions by individuals influential in the field who hold to a variety of interpretations on the question of what *is* the international system.

Robert Keohane and Joseph Nye authored an influential study in the late 1970s which looked at the relationships between two critical variables: power and interdependence. Their intention was to take a broad overview of the international system and to scrutinize recent changes, thereby attempting to draw out patterns of regularized state behavior.[68] The terms *dependence* and *interdependence* are very useful components for an effectual grasp of many of the dynamics present within the international system.

> In common parlance, *dependence* means a state of being determined or significantly affected by external forces. *Interdependence*, most simply defined, means *mutual* dependence. Interdependence in world politics refers to situations characterized by reciprocal effects among countries or among actors in different countries.[69]

Leading indicators for the effect studied by Keohane and Nye "... result from international transactions—flows of money, goods, and messages across international boundaries. Such transactions have increased..." significantly since 1945. The post-war economic boom (and the periodic recessions), the growth in the role of science and technology in everyday life and the increase in world travel have added to the expansion of these "flows" which were identified. There are important differences, however, separating a greater "connectedness" between these international transactions and a greater interdependence. For the two authors

> ... [I]nterconnectedness is not the same thing as interdependence. The effects of transactions on interdependence will depend on the constraints, or costs, associated with them... Where there are reciprocal (although not necessarily symmetrical) costly effects of transactions, there is interdependence. Where interactions do not have significant costly effects, there is simply interconnectedness. The distinction is vital ... to understand the *politics* of interdependence.[70]

These two authors posit an interesting relationship between the reality of nuclear weapons and the continuing importance of naval forces as a tool of power projection for the nation-state. There have been many changes in international relations within the last 45 years, but one thing the world has *not* seen is the elimination of force and threat from the "repertoire" of interstate behavior. Patterns of action have been modified by a variety of factors, and continue to be modified, but many bases of human and political interaction have remained the same since such questions were first raised in the West by the ancient Greek philosophers: What is the human being? And, what is our relationship with others and the self? Keohane and Nye do not attempt to reconcile this debate within *Power and Interdependence*, but they do hold that the relevance of the traditional connection between military and political power–suitably updated–is particularly pointed when the focus of attention is directed at the changing behavior of the United States and the then Soviet Union (when that nation existed) on the multilateral level.

The author's book also raises important issues in the question of the traditional limits of state power and the global commons. Historically, as was discussed earlier in the Preface, the sovereign power of the state was exercised on *terra firma*. This situation remained essentially stable over time, but important changes were also slowly occurring over the course of many centuries as peoples within the Mediterranean basin expanded their patterns of exploration and settlement. Similar patterns were replicated tens of centuries later when humankind began a "second" age of exploration in the 1500s, this time spreading the activity over the entire planet. Keohane and Nye do not range this far in their examination, but the expansion of a state's power over activities not on land—primarily driven by questions of war, trade and commerce (and later the possible exploitation of economic resources from the deep sea-bed because of technological advances discovered in the 20th century)—is an important issue to which they devote considerable attention.[71]

Power and Interdependence is a transitional work which seeks to bridge the differences between the Realist school of international relations and the underlying conversion introduced into that system by the growth of non-state actors (i.e., non-governmental organizations) since 1945. The appellation Neo-Realism can be properly applied to this effort.

Mastanduno, Lake and Ikenberry have addressed the question of linkages between domestic and international politics.[72] Their task is to investigate "the growing interaction of international and domestic politics ... by utilizing conceptual frameworks that address not only the impact of one on the other, but the interplay between..." the two factors as well.[73] Their effort is a preliminary one—an exposition on where further research should lead. As the conclusion notes, "An adequate empirical test of the propositions generated by this approach is beyond the scope of this present paper. The illustrations provided, however, should be sufficient to indicate the potential utility of the approach."[74]

Two schools of thought are present within the Realist paradigm—the Classical and Structural models. The three essential arguments which are present within the Classical approach pertain directly on the use or abuse of ocean resources.

Conclusion

This chapter has examined the components of the international system, focusing on the contributions of both historical and contemporary theorists. It has discussed the organization of the current state system, which is based on sovereignty and territorially defined nation-states, and has addressed some of the more important changes affecting this state system. It has not, by design, attempted to achieve a complete or full rendering of current international politics or the contributions of international relations scholars. The purpose has been, rather, to provide a (hopefully) thorough grounding in the theoretical issues that will make clear the points to be raised in the following chapters. Those chapters will examine the actual programs and activities of intergovernmental organizations and international non-governmental organizations, beginning first with a review of the physical state of the ocean in selected areas. The crisis, and the potential for devastating ecological consequences, have profound implications both for peoples and sovereign states.

Chapter 2 Notes

[1]I define the state system as the current pattern of organization on the international level, i.e. made up of sovereign states, territorially based, impermeable, with no recognized higher legal authority than themselves.

[2]See Maurice Keens-Soper, "The Practice of a States-System," in Michael Donelon, *The Reason of States, A Study in International Political Theory* (London: George Allen & Unwin, 1978), p. 31. Also, see Martin Wight, *Systems of States* edited by H. Bull (Leicester, England: The University Press, 1977).

[3]See Michael Donelon, op. cit., p. 14 (Introduction).

[4]Ibid.

[5]I define global politics as a systemic level of analysis above the systemic study or approach of international politics. It is, also, in a process of development; a realization of the existence and urgency of *The Global Problematique*. This term was first defined by Donella Meadows et al., in *The Limits to Growth* (New York: Universe Books, 1972); quoted in Dennis Pirages, *Global Technopolitics, The International Politics of Technology and Resources* (Pacific Grove, CA: Brooks/Cole Publishing Company, 1989), p. 50, footnote. 3. As Pirages notes,

"The term ... is used to describe an interlocking set of environmental, population and institutional problems." International politics is organized primarily, although certainly not exclusively [cf. international non-governmental organizations (INGOs) defined by Werner J. Feld and Robert S. Jordan, with Leon Hurwitz in *International Organizations, A Comparative Approach*, 2nd ed. (New York: Praeger Publishers, 1988), p. 7], through the systemic activities of sovereign nation-states. In addition, global politics is to be distinguished from earlier and more primitive stages of human social evolution where the primary locus of political activity outside the still existent nation-state is confined to "international politics." Global politics is a stage of human growth and adaptation wherein common problems of humanity, inherently transnational in character (global environmental protection, global economic security and growth, and global physical security from war and other forms of inter-community violence) are studied through a prism of joint and coordinated approaches not just of nations but of an inherently strengthened and transformed international system. This new global politics is a result of transition—a transition that can be studied by a utilization of more complex indices of political behavior than that heretofore dominant within one field of international relations.

[6]Alan James, *Sovereign Statehood, The Basis of International Society* (London: Allen & Unwin, 1986), p. 1.

[7]Ibid., pp. 3-4.

[8]Ibid.

[9]This section is from Alan James, op. cit., Ibid.

[10]This section is also drawn from Alan James, op. cit., p. 13. Refer, in addition, to the work of Falk on Westphalia and the Charter in Richard Falk, Kratochwil Friedrich and Saul H. Mendlovitz, *International Law: A Contemporary Perspective* (Boulder, CO: Westview Press, 1985), p. 121. F. W. Hinsley in *Sovereignty*, 2nd ed. (Cambridge: Cambridge University Press, 1986), p. 159 notes: "When the theory of sovereignty within the community was formulated in Western Europe in early modern times, the difficulty of applying it to the relations between communities proved to be almost as great as ... arriving at the sovereignty notion within the separate community... The basic reason ... is obvious enough. It occurred because ... [individuals] had to overcome immense obstacles before they could conceive of the world they inhabited as being a world composed of separate political communities."

[11]Permanent Court of Arbitration (1928), Reports of International Arbital Awards, vol. 2, p. 829. Quoted in Ingrid Detter DeLupis, *International Law and the Independent State* (Aldershot, England: Gower Publishing Co., Ltd. 1987), p. 4.

[12]Alan James, op. cit., p. 6.

[13]Ibid., pp. 23-24.

[14]Ibid., p. 25.

[15]This raises the fascinating question whether states sometime in the future would cede some of their sovereignty in a real and meaningful sense to a global organization because it was felt, at least regarding the issues involved, that such a new multilateral approach was more likely to avert imminent catastrophe or current system breakdown, or both. Concomitantly, one can ask whether international integration efforts currently underway need to reach a point "beyond recall" before the state system as we know it would lose its predominance as the most powerful form of political organization? Alan James, op. cit., raises this point, p. 252.

[16]See the following publications: Raphael Israeli, ed., *PLO in Lebanon, Selected Documents* (London: Weidenfeld and Nicolson, 1983); Aaron David Miller, *The PLO and the Politics of Survival* (New York: Praeger Publishers, 1983); and Aryeh Y. Yodfat and Yuval Arnon-Ohanna, *PLO, Strategy and Tactics* (New York: St. Martin's Press, 1981).

[17]DeLupis, op. cit., p. 4.

[18]Power is a difficult concept to describe, although many have adequately done so. See particularly Hans J. Morgenthau, *Politics Among Nations, The Struggle for Power and Peace*, 3rd ed. (New York: Alfred A. Knopf, 1963), especially Chapter 3 "Political Power," pp. 27-37.

[19]F. H. Hinsley, *Sovereignty*, op. cit., p. 1.

[20]I am indebted to Clive Archer of the University of Aberdeen (Scotland) for this perceptive observation.

[21]I am again grateful to Clive Archer for clarification on the issue of the history of ocean regimes. His ideas were critical in the development of this section. Also, see Robert O. Keohane and Joseph S. Nye, *Power and Interdependence, World Politics in Transition* (Boston and Toronto: Little, Brown and Co., 1977), pp. 90-92.

[22]See Lynn H. Miller, *Global Order, Values and Power in International Politics*, 2nd ed. (Boulder, CO: Westview Press, 1990), p. 215 and Clyde Sanger, *Ordering the Oceans, The Making of the Law of the Sea* (Toronto and Buffalo: University of Toronto Press, 1987), pp. 11-13.

[23]Clyde Sanger, op. cit., pp. 19-20. See also "The sea bed, international area." *UN Chronicle*, Vol. XIX, No. 6, June 1982. Quoted in Jacques G. Richardson, ed., *Managing the Ocean, Resources, Research, Law* (Mt. Airy, MD: Lomond Publications, Inc., 1985), pp. 271-279; and Rene Jean Dupuy, "The Convention on the Law of the Sea and the new international economic order," in Jacques G. Richardson, ed., op. cit., pp. 281-295.

[24]For a discussion of the New International Economic Order see Robert S. Jordan, "Why an NIEO: The View from the Third World," in Harold Jacobson and Dusan Sidjanski, eds., *The Emerging International Economic Order: Dynamic Processes, Constraints and Opportunities* (Beverly Hills, CA: Sage, 1982). In addition, refer to Robert W. Gregg, "The Politics of International Economic

Cooperation and Development," in Lawrence S. Finkelstein, ed., *Politics in the United Nations System* (Durham, NC and London: Duke University Press, 1988), pp. 106-147, especially pp. 108-115 and 124-142. Also confer Lawrence S. Finkelstein, "The Politics of Value Allocation in the UN System," in Lawrence S. Finkelstein, ed., op. cit., pp. 1-40; and Robert L. Friedheim, "Value Allocation and North-South Conflict in the Third United Nations Law of the Sea Conference," also in Lawrence S. Finkelstein, ed., op. cit., pp. 175-213.

[25]See Leigh S. Ratiner, "The Law of the Sea: a Crossroads for American Foreign Policy." *Foreign Affairs* (Summer 1982), p. 1008. Quoted in Clyde Sanger, op. cit., pp. 49-50.

[26]See M.C.W. Pinto, "Emerging concepts of the Law of the Sea: some social and cultural impacts," in Jacques G. Richardson, op. cit., pp. 298-309.

[27]Clyde Sanger, op. cit., pp. 194-211.

[28]See Patricia Birnie, "The role of international law in solving certain environmental conflicts," in John E. Carroll, ed., *International Environmental Diplomacy* (Cambridge: Cambridge University Press, 1988), pp. 102-104; and Robert E. Riggs and Jack C. Plano, *The United Nations, International Organization and World Politics* (Chicago: The Dorsey Press, 1988), pp. 254-258.

[29]See Robert B. Abel, "The History of the United States Ocean Policy Program," in Francis W. Hoole, Robert L. Friedheim and Timothy M. Hennessey, eds., *Making Ocean Policy, The Politics of Government Organization and Management* (Boulder, CO: Westview Press, 1981), pp. 3-48.

[30]See Werner J. Feld and Robert S. Jordan with Leon Hurwitz, *International Organizations, A Comparative Approach*, 2nd ed. (New York: Praeger Publishers, 1988), p. 248.

[31]See Hedley Bull and Adam Watson, eds., *The Expansion of International Society* (Oxford: Clarendon Press, 1984), p. 9.

[32]Hedley Bull, *The Anarchical Society, A Study of Order in World Politics* (New York: Columbia University Press, 1977), pp. 8-13. For Bull, "[a] society of states (or international society) exists when a group of states, conscious of certain common interests and common values, form a society in the sense that they conceive themselves to be bound by a common set of rules in their relations with one another, and share in the working of common institutions." He further notes, "If states today form an international society ..., this is because, recognising certain common interests and perhaps some common values, they regard themselves as bound by certain rules in their dealing with one another ...; claims to independence ... honour agreements ... subject to certain limitations in exercising force against one another" (p. 13). International society is discussed as part of a troika of definitions, the other two being "system of states" and "international system." Again, it is best to quote directly from Bull. "*A system of states* ... is found when two or more states have sufficient impact on one another's decisions, to cause them

to behave—at least in some measure—as parts of a whole" (pp. 9-10). Raymond Aron defined international system (here used interchangeably with Bull's "system of states") as "... the ensemble constituted by political units that maintain regular relations with each other and that are all capable of being implicated in a generalized war." See his *Peace and War, A Theory of International Relations*, trans. Richard Howard and Annette Baker Fox (Garden City, NY: Doubleday and Company, Inc., 1966), p. 94. Bull in *The Anarchical Society* discussed international system: "The term 'international system' has been a fashionable one among recent students of international relations, principally as a consequence of the writings of Morton A. Kaplan... Kaplan's use of the term is not unlike that employed here, but what distinguished ... [his] work is the attempt to use the concept of a system to explain and predict international behavior, especially by treating international systems as a particular kind of 'system of action.'" See Morton A. Kaplan, *System and Process in International Politics* (New York: Wiley, 1957), p. 4. Quoted in Bull, op. cit., p. 12.

[33]This paragraph is taken from the work of Peter Gellman, "Hans J. Morgenthau and the legacy of political realism." *Review of International Studies* (October 1988), p. 255.

[34]Ibid.

[35]See Hedley Bull in Hedley Bull and Adam Watson, eds., op. cit., pp. 224-226 and K. J. Holsti, *International Politics, A Framework for Analysis*, second edition, (Englewood Cliffs, NJ: Prentice-Hall, Inc., 1972), pp. 67-71.

[36]See Volker Rittberger, ed., *Science and Technology in a Changing International Order, The United Nations Conference on Science and Technology for Development* (Boulder, CO: Westview Press, 1982).

[37]See footnote 36 (emphasis added).

[38]This holds true for other European empires as well, such as the Dutch withdrawal and granting of independence to Indonesia. The Portuguese example is a separate case because, there, the colonial empire in Mozambique and Angola did not collapse until the downfall of the Salazaar dictatorship in 1975. In addition, see Lynn H. Miller, *Global Order, Values and Power in International Politics*, second edition (Boulder, CO: Westview Press, 1990), pp. 146-151.

[39]See footnote 36. At the time of this writing (Winter 1990) a profound transformation in global politics is occurring from what can be called the Democratic Revolution of 1989-1990 in Eastern Europe, particularly Germany, and the collapse of the Leninist system in the Soviet Union. This does not negate Bull's analysis that the actions of Soviet policy were a significant contributing cause to global instability. See R. Judson Mitchell, *Ideology of a Superpower, Contemporary Soviet Doctrine on International Relations* (Stanford: Hoover Institution Press, 1982), especially Chapter 1 "Introduction," and Chapter 6 "Proletarian Internationalism: The USSR and World Revolution."

[40]There have been advantages to this, chief among them the ending of colonialism. Perhaps this is the only truly significant and positive contribution Marxist-Leninist philosophy has made to world culture and civilization.

[41]Christer Jonsson, *Superpower, Comparing American and Soviet Foreign Policy* (London: Frances Pinter Publishers, 1984).

[42]See David H. Blake and Robert S. Walters, *The Politics of Global Economic Relations*, 3rd ed. (Englewood Cliffs, NJ: Prentice-Hall, Inc., 1987) and Richard Rosecrance, *The Rise of the Trading State, Commerce and Conquest in the Modern World* (New York: Basic Books, Inc., 1986).

[43]This section is drawn from K. J. Holsti, *International Politics, A Framework for Analysis*, 2nd ed., op. cit., pp. 66-71.

[44]The problems transferred to the international system from sovereign states which have nationalities within that do not wish to be there are obvious. Parenthetically, of course, this is one of the main problems associated with the principle of political organization.

[45]Holsti, op. cit., p. 69.

[46]See his *Culture and Political Change* (New Brunswick, NJ: Transaction Books, 1983).

[47]Ibid., p. 2.

[48]Ibid.

[49]Ibid., p. 3.

[50]K. J. Holsti, op. cit., p. 71.

[51]The December 1990 all-German elections is the capstone to the political collapse of the Soviet Union in its once heretofore allied Eastern Europe. See *The New York Times*, 3 December 1990, p. 1.

[52]The events of June 1989 and the brutal crushing of the pro-democracy movement in Beijing by Communist authorities illustrates the point.

[53]See *The New York Times*, 14 February 1990, p. 1.

[54]See Werner J. Feld and Robert S. Jordan, op. cit., p. 248.

[55]See E. H. Carr, *The Twenty Years' Crisis, 1919-1939, An Introduction to the Study of International Relations* (London: Macmillan and Co., Limited, 1940); Walter Lippmann, *U.S. Foreign Policy* (Boston: Little, Brown and Company, 1943); Hans J. Morgenthau, Politics Among Nations, *The Struggle for Power and Peace*, 3rd ed., op cit., John Lewis Gaddis, *The United States and the Origins of the Cold War, 1941-1947* (New York and London: Columbia University Press, 1972); and John Spanier, *American Foreign Policy Since World War II*, 10th ed. (New York: Holt, Rinehart and Winston, 1985).

[56]Ernst B. Haas, *Beyond the Nation-State, Functionalism and the International Organization* (Stanford: Stanford University Press, 1964).

[57]Ibid., p. 6.

[58]Ibid., p. 497.

[59]See Inis L. Claude, Jr., *Swords into Plowshares, The Problems and Progress of International Organization*, 4th ed., (New York: Random House, Inc., 1971).

[60]This paragraph and the next is drawn from Inis L. Claude, Jr., *Swords into Plowshares, The Problems and Progress of International Organization*, op. cit., pp. 447-448.

[61]The 1983 invasion of Grenada by American armed forces is another case in point.

[62]See Hans J. Morgenthau, *Politics Among Nations, The Struggle for Power and Peace*, 3rd ed., op. cit., pp. 4-15 for a complete elucidation.

[63]See Peter Gellman, "Hans J. Morgenthau and the legacy of political realism," op. cit., p. 248 in Hans J. Morgenthau, *Scientific Man Versus Power Politics* (Chicago, 1946), pp. 37-40.

[64]Peter Gellman, "Hans J. Morgenthau and the legacy of political realism," op. cit., p. 254 in Hans J. Morgenthau, "The Problem of the National Interest" in Morgenthau, *Politics in the Twentieth Century* (Chicago, 1962), vol. 1, p. 93.

[65]Raymond Aron, *Peace and War, A Theory of International Relations*, trans. Richard Howard and Annette Baker Fox (Garden City, NY: Doubleday and Company, Inc., 1966), p. xvi.

[66]Ibid., p. 9.

[67]Ibid., p. 10.

[68]Robert O. Keohane and Joseph S. Nye, *Power and Interdependence, World Politics in Transition* (Boston and Toronto: Little, Brown and Co., 1977).

[69]Ibid., pp. 8-9. This section is drawn from these pages.

[70]Ibid., p. 9.

[71]See Robert O. Keohane and Joseph S. Nye, op. cit., especially "Part II, Regime Change in Oceans and Money," pp. 63-162, especially "The Oceans Issue Area" in Chapter 4, "The Politics of Oceans and Money: Historical Overview," pp. 86-98. In this section, the authors posit three periods of historical coalescence with regard to international regimes and the oceans policy issue system. Period I covers the years 1920 through 1945, and the "regime situation" is characterized as a "Free seas regime." Period II, from 1946 through 1966, developed into and is described as a "Strong quasi regime." Finally, Period III, from 1967 through 1975 is designated a "Weak quasi regime" (p. 92). International regimes are defined as the product of the interdependence that "affects world politics and the behavior of states, but [their definition also notes that] governmental actions also influence patterns of interdependence ... [In addition by] creating or accepting procedures, rules, or institutions for certain kinds of activity, governments regulate and control transnational and interstate relations." It is such "governing arrangements" that are defined by Keohane and Nye as "international regimes" (p. 5).

[72]See Michael Mastanduno, David A. Lake and G. John Ikenberry, "Towards a Realist Theory of State Action," *International Studies Quarterly* (December 1989).

[73]Ibid., pp. 457-459.

[74]Ibid., p. 471.

3

The Physical State of the Ocean and Pollution: The Crisis

Part I

Introduction

The dominant fact of life in the biosphere of Planet Earth is the existence and sheer size of the many open bodies of water properly called the Global Ocean. The overwhelming predominance of water on the planet is often overlooked because of its very familiarity. Traditionally, we have looked upon this ocean not as one but as five distinct and very large entities—the Antarctic, the Arctic, the Atlantic, the Indian and the Pacific. In point of fact, however, such a classification scheme is outmoded; and of extremely limited analytical value. All of the "oceans" are in fact a single entity. All sustain and influence each other in a myriad of atmospheric, biological, chemical and physical interactions. To be sure, continued use of these terms is helpful for descriptive, historical, literary or sentimental purposes. But they are not of significant scientific use except as regards the fact of their existence as parts to a greater whole. This is a point of epistemology.

It is not my intention here to confuse the contributions of those studies which take a part (a particular "ocean," say) of the Global Ocean and study, for instance, flux patterns in order to determine heat transfers or chemical compositions of coastal waters and its effects on phytoplankton reproduction for marine fisheries. Of course, such work is a major and significant contribution to the understanding of the marine environment (major planetary studies now underway by intergovernmental organizations will be examined in Chapter 4). Rather, the point concerns a change in understanding concepts relating to the unity of oceanographic studies and their applicability to problems of government and international relations. There is no use in adhering to historical definitions which reflect a limited vision or scope in matters respecting the study of ocean problems. When various nations and peoples were in the age of exploration and discovery (roughly five hundred years ago) these individual oceans assumed a special importance in world history. This process reflected a stage in the development of a global society. Now that the scientific bases of oceanography and related sciences are better understood, it is time for the changes introduced within recently published texts to make their way into wider segments of society and political discussion.

Oceanography[1]

Causal relationships in the physical and natural sciences do not receive nearly enough attention in the international relations literature. The result is a blinded or incomplete analysis not only of the relations among states, particularly the non-military/security causes for many conflicts, but of the more prosaic workings of the international system itself—why certain issues are becoming more important than others. Such deficiencies can in part be remedied by including non-traditional sources in the work of international organization studies. As a modest beginning, such an interdisciplinary approach is utilized in this work.

The distribution of land mass and ocean on Planet Earth is very uneven. All of the continents cover only 29.2% of the planet's surface. This distribution divides the ocean into "three basins" which have the well-known designations of the Atlantic, Indian and Pacific. As I have already mentioned, water itself covers more than 70% of the earth's surface. Gross notes that these "ocean basins ... can be viewed as northward-projecting gulfs of the world-circling ocean around Antarctica, the [true] Southern Hemisphere." In the Antarctic, this ocean accounts for 80.9% of the total area.[2] Ours is overwhelmingly a water planet, but humankind has, of course, primarily evolved in a terrestrial environment. In this century, and essentially in the last four or five decades, serious and sustained attention has been directed towards the problems of the ocean. The intensity and scope of these very recent discoveries have characterized the tremendous growth in knowledge between this century and ones previous; when, of course, studies in oceanography and exploration were also carried out.

In 1972 the first widely noticed and distributed work in the West on environmental issues was one by Barbara Ward and Rene Dubos. Their *Only One Earth* was distilled as a "working document" from wide-ranging consultations with environmental experts all over the globe. Delegates to the then up-coming Stockholm conference on the Human Environment (officially called the United Nations Conference on the Human Environment) utilized the report during their deliberations in the Swedish capital. Perspicaciously they noted

> It was in the oceans, after the secular downpour of the early rains, that life first began to stir, shielded by the waters from the sun's irrestible radiation. It was from the oceans that plants and animals emerged to colonize the land surface of the planet. It is the oceans today that provide the water vapor

> which, drawn up by the sun, falls upon the earth–in harvest–bringing life-sustaining rain. Ocean water is our planet's filtering system where all debris, both mineral and biological is dissolved, decomposed, and transformed into life-supporting substances... It is a major provider of the oxygen released by its phytoplankton for the benefit of all the species... Without water's special qualities for holding heat, much of the earth would be uninhabitable. The oceans are the coolants of the tropics, the bringers of warm currents to cold regions, the universal moderators of temperature throughout the globe.[3]

Harold and Margaret Sprout were also pioneers in pointing out to a largely unknowing world the consequences of the destruction and neglect of the environment.[4] Conflict arising from such ignored and neglected issues are now being addressed within the field of international relations. *Bordering on Trouble: Resources and Politics in Latin America* is one fairly recent example.[5] This edited collection is an examination of political, economic and social indicators which chart factors of instability in Latin America. The thrust of the book's argument revolves around the distortion and mis-application of environmental resources.

Because of its astonishing size, the ocean has historically been viewed as beyond the influence of human beings. Such views are, of course, undergoing transformation; but the historical power and lasting impact of such concepts are powerful and not easily supplanted. They do more than just "linger." Such views powerfully and subtly influence governments and are still very widely held within many sectors of society in the West (as examples below will show).

Ward and Dubos have commented on this problem of the ignoring of social and scientific reality in this way.

> [T]he ocean system must seem, at first thought, to be infinitely beyond the reach of man's [sic] puny influence... In fact, men [and women] are still under the strong influence of the medieval concept of an endless ocean. We all tend to feel that once a polluted river empties into the open sea, once we conduct sewer systems far enough away from land, all industrial and urban discharges will disappear somehow into blue space beyond the horizon, as if we pipe it away from our own planet. We seem in this conception of the ocean to forget ... *that the world is round and without edges.*

Advances in satellite technology have increased tremendously since the 1970s. These improvements have made it possible to much more accurately chart many of the "short-term processes affecting open-ocean currents."[6] In turn, this development has added to the body of knowledge in oceanography revealing more detailed patterns of ocean current circulation which is an important component for understanding dispersal and transportation patterns of marine pollution. The "short-term processes ... [of] open-ocean currents" provide researchers with a more accurate and "real-time" image of ocean events. Such knowledge is necessary if scientists are to know of immediate as well as long-term trends in ocean science.

Utilizing reports from the United Nations and associated bodies or agencies, composed of scientific and technical experts, this chapter analyzes in Part II five major types of pollution in light of their impact on the international system. First, however, some additional points need to be made concerning the science of oceanography. This is needed in order to understand the probable repercussions for the international system of massive social and economic dislocation which will be caused by severe degradation of the marine environment.

The most widely quoted definition for pollution is the one developed and used by the Joint Group of Experts on the Scientific Aspects of Marine Pollution (GESAMP). GESAMP is a body of specialists comprising members of several United Nations agencies.[7] The definition of marine pollution is as follows:

> Pollution means the introduction by man, [sic] directly or indirectly, of substances or energy into the marine environment (including estuaries) which results in such deleterious effects as harm to living resources, hazards to human health, hindrance to marine activities including fishing, impairment of quality for use of sea water and reduction of amenities.

"The GESAMP definition was endorsed by the Stockholm Conference ... in 1972 ...," having first been proposed in 1969.[8]

The interstices of pollution and international politics are made particularly evident on the shoreline of littoral states. There, where land, sea and air all meet "... is the most dynamic part of the ocean."[9] It is also the location where often the most intractable problems of pollution interact with the exigencies of contemporary international relations.

One reason the shoreline is so important, and why it contains such continuing potential for controversy, conflict and disaccord, is that "processes affecting these shores act for periods of time ranging from a few seconds to tens of thousands of years."[10] Historically, international relations have been anything but organized on this type of time scale. Issues of contention between nation-states have almost always been organized around time periods far less than even centuries. Indeed, wars have usually lasted only several years. The incessant activity associated with coastal processes makes for a constant interaction on the frontiers of nation-states. When problems of externally induced pollution interact with shorelines—such as oil spills—the political repercussions are quick to emerge.[11]

A longer-term problem for international relations is produced by the issue of climactic change and global warming. This issue will receive more detailed treatment in the following section on types of pollution; suffice to say for now, however, the ultimate impact of this development could be as serious as any humankind has faced.

> The most profound influence on coasts as we now see them was the repeated advance and retreat of continental glaciers and resulting sea level changes... When sea level stood at its lowest, about 18,000 years ago, the shore was near the present continental shelf break. As the glaciers waned and water was released, the shoreline moved back ... until it reached its present level about 3,000 years ago. If the remaining ice sheets (Antarctica, Greenland) melt, the sea surface may eventually stand as high as 50 meters [164 feet] above its present level. Today's shoreline would be submerged...[12]

Put another less dramatic way, expected results from a rise in sea-level would include: "inundation of low coastal lands, relocation or destruction of coastal wetlands, shoreline erosion and beach loss, exacerbated exposure to storm surge and flooding, and increased salinity of rivers, bays and aquifers." Ice sheets would not have to completely melt for catastrophic results to occur. A relatively and very small loss of total mass of frozen ice would induce widespread global coastal destruction, massive population transfers, the total inundation of some small island nation-states (the Maldives, for instance) and raise the spectre of military conflict because of environmental deterioration.[13]

"There are three major depth zones in the ocean: surface, pycnocline (the density boundary between the surface and the deep ocean), and deep zones..."[14] Temperature is a critical component in the healthy functioning of this system. The continual effects of human activity, particularly over the past 100 years, however, may be significantly changing the temperature of the water. An experiment is now in the planning stages which hopes to begin operation in 1992 which will utilize an "... elaborate experiment to use sound waves to measure the temperature of all of the world's oceans simultaneously..."[15] The goal of this work is to "determine whether the greenhouse effect has caused the temperatures of the oceans to rise as climate modellers predict." A successful survey may indicate how the ocean is acting as a "global sink" for increased heat generated by carbon monoxide and other atmospheric pollutants. The surface zone of the ocean is in the most direct and immediate contact with the atmosphere. "It undergoes substantial changes because of seasonal variability in local water budgets (evaporation, precipitation) and heat budgets (incoming solar radiation, evaporation)."[16] In many ways, the surface zone is the most active of the three layers because it is "mixed" with so many other factors; including, wind, waves and water from deeper layers which are cooled by surface waters "... in the open ocean."[17] Other ocean layers also interact with temperature and atmospheric variables, but to a lesser extent than the pycnocline.[18]

Mariculture is a major potential issue for conflict in international relations and is directly related to the salient issues of coastal areas and fisheries production. This is a near perfect example of the transformation in international politics which has characterized the emergence of traditionally defined "low politics" controversies into questions of "high politics."[19]

Gross states that "mariculture, or aquaculture, ... is a means of obtaining more food from the ocean than the normal fishing and gathering activities. It usually involves modifying and controlling benthic ["...that portion of the marine environment inhabited by marine organisms that live permanently in or on the bottom"] environments."[20] Certain very necessary and sensitive biological and oceanographic processes are required for a successful mariculture industry. For photosynthesis to occur, several key elements must be present. Roger Revelle has written that

> The productivity of marine plants, like that of plants on land, depends on the amount of sunlight and on the availability of nutrients, including nitrogen compounds, phosphorus, potassium and other mineral substances. Photosynthesis is possible only in the upper ocean layers where bright sunlight penetrates.[21]

Up-welling—the process whereby "water rises from a lower to a higher depth, usually caused by divergence and offshore currents"[22]—photosynthesis, and productivity of kelp fields (discussed below) are impacted directly by ocean pollution. The "introduction of substances or energy into the marine environment..." can alter the delicate balance upon which these enterprises can flourish. Not only are the "coastal upwelling regions ... the most productive areas in the ocean," but the coastal ocean itself lying on the margin of the continental land mass often encompasses the most complex and intricate patterns of life. "The shallow bottom and usually complex shoreline make the coastal ocean far more complicated and difficult to understand than the open ocean."[23]

Kelp is substantially sized brown benthic algae. It "grow[s] in shallow subtidal rocky areas in subtropical to subpolar coastal oceans."[24] It is a potentially major, yet largely untapped, food source for world population (the difficulty of its introduction into diets because of perceived cultural disadvantages notwith-standing). As with all productive food areas of the ocean, its sensitivity to the introduction of pollutants is high. The growth parameters by temperature are also quite strict—kelp can only grow in waters "cooler than 20 degrees Celsius that are nutrient-rich" [68 degrees F].[25]

The few issues discussed here are a small sample, selectively chosen, to illustrate the connections between oceanography and international politics. These themes will be further explored in the following chapter (on IGOs), as well as the subsequent one (on INGOs). The preceding discussion of oceanographic issues is a useful tie-in to the next section of this chapter which begins below.

The Interrelatedness of Biological, Chemical and Physical Processes

The interrelatedness and unity of biological, chemical and physical processes nowhere makes itself more presciently visible than in the realm of ocean pollution. The ocean is a "sink" for all of the activities of humankind, the depository of its effluents and detritus. As the explorer and oceanographer Jacques-Yves Cousteau has observed

> In publication, in conference, in international units the matters are generally divided into air pollution, land pollution and water pollution. In fact, there is only one pollution because every single thing, every chemical whether in the air or on land will end up in the ocean.[26]

Cousteau's conclusion is supported by the findings of various scientific bodies in this section and discussed below. His conclusion is also supported by Pirages who has analyzed the impact of industrialization on the "global environment."[27]

GESAMP also studies the interrelatedness and unity of biological, chemical and physical processes of the ocean. In 1986 the group released an important report which provides "... guidelines for the assessment of the impact of potentially harmful substances released into the marine environment."[28] This is a comprehensive and leading attempt to introduce practical and meaningful steps to measure, in stages if necessary, the Environmental Capacity of the marine environment ("... defined as a property of the environment, a measurement of its ability to accommodate a particular activity or rate of an activity, such as the discharge of contaminants, without unacceptable impact").[29] The report is not a "how to" effort on the ways and means of gathering the necessary basic data for the implementation of the guidelines. As the authors rightly state, "To have done so would simply have duplicated material which is already available in the open literature and therefore accessible to those persons who will be brought into advise or otherwise provide expert opinion on any project."[30] What it does do is concentrate "on describing the parameters and processes which have to be taken into account in the assessment of the impact of pollutants on marine organisms, ecosystems, amenities and human health, as a consequence of any discharges to the marine environment."[31]

Officially GESAMP is not a political action group of the United Nations System in business to recommend specific policy solutions to national governments for correcting the problems of the marine environment. As its' brief clearly states, and as many of the reports note in a preface

> According to its present terms of reference, the functions of GESAMP are: … to provide advice relating to the *scientific aspects* of marine pollution; and to prepare periodic reviews of the state of the marine environment as regards marine pollution and to identify problem areas requiring special attention.[32]

The above study which addresses the concept of Environmental Capacity is a proposal not only to guide future research, but also is part of an on-going process seeking to contribute to a more complete and accurate picture on the state of the global ocean and the marine environment in general. The authors lay out the underlying assumptions to their report that: "... (1) a certain level of any contaminant will not produce any unacceptable effect on the marine environment or its various uses; (2) the environment consequently has a finite capacity to accommodate wastes; [and] (3) such capacity can be quantified."[33] Specific recommendations for individual marine sites are not possible because the problems and solutions differ with each particular circumstance. However, studies which analyze variable objectives, set targets and pathways are useful in addressing the myriad of problems associated with the marine environment.[34]

Nurmi has noted the "considerable advances" that were made, particularly during the 1970s, in the field of oceanography. A greater understanding of "the physical and chemical properties of the ocean, and of the circulation of their waters …" was achieved in this decade by scientists and technicians alike.[35] Policy makers and analysts, both on the national and international levels, were thus presented with significantly more accurate and complete analyses of ocean processes. However, such gains in knowledge, while significant, were still very limited. Many areas of this increased understanding were left very incomplete. For instance, "… the monitoring of water quality and of sediment and living organisms is in general quite recent and is not yet sufficiently developed to show whether pollution is increasing or decreasing, except in specific cases where vigorous pollution control measures have been applied."[36] The implications for a political analysis of ocean pollution and the international and national efforts to address the issue are fairly obvious. Not only are the physical and natural scientists (oceanographers, biologists and the like) constrained by the limited scope of their efforts (in the sense of being able to broadly understand, describe and predict the extensive processes operative within the ocean);—this point is quite obvious and plagues many of the sciences—but decision makers on the state level often ignore the findings and recommendations of the "experts" they specifically find and charge to study such problems.

It is an axiom to state that in the circumstances described above humankind is impacting upon the environment in ways that are far from being fully or even, in some cases, partially understood. The impact of this human intervention is problematic. History will judge whether the efforts to redress were sufficient. The medieval historian C. H. Lawrence noted "History is a language of explanation."[37] The future cannot be explained until it develops. However, the task of social scientists in this matter is to study the problem as thoroughly as the current state of knowledge permits and to draw conclusions which then can be used to design appropriate remedies.

One final point needs to be made before proceeding to the source pollution question. It may belabor the point to briefly lay out some of the possibly deleterious effects of pollution on the ocean in the context of this study, but again I shall quote Nurmi.

> Polluting substances may cause changes in the physical-chemical environment, e.g. in temperature, salinity, pH, gas content, chemistry, colour or turbidity of the water, which in turn may influence the marine life in different ways. They may also cause direct damage to organisms, e.g. on physiological processes, including alimentation and reproduction, after having reached a certain threshold concentration.

It can be stated equally without qualifying comments that pollution has increased to dramatic proportions over the planet. "Pollutants have spread enormously, especially since the early 1950s, and no part of the seas, even at great depths, seems now to be completely free from them. Nobody knows how many pollutants there are, but already the pertinent new chemical compounds can be counted in tens of thousands."[38] Nuri's research was published in 1988. Such conclusions are now even more pertinent.

Finally, there are many difficulties involved with the measurement of marine pollution.[39] The methods used to calculate the levels of impact or degradation need to be "harmonize[d]," and "available data" is at many times meager or narrow. Findings are in many cases difficult to interpret, and there has not been a uniformity of effort in seeking information from all available regions. And, Nurmi has noted that "while the general characteristics of the world oceanic circulations, seawater composition and biological production have been broadly established, the *monitoring* of pollutants and ecosystems has been highly localized and time series data are lacking for most cases."

Part II

The following section will address the five major types of pollution altering the ocean environment; (1) run-off from land, (2) transboundary transportation of toxics, (3) oil, (4) nuclear waste and (5) atmospheric. Concrete localized expression to the "dispersal paradox" offered in the work of Pirages is a major theme.

Run-off from land sources

It has long been noted "...that the greater part of marine pollution is caused by landbased activities. Their exact contribution to the total load of marine pollution remains difficult to determine but a proportion from 50 to 90% has been estimated."[40] The reasons for this are quite clear and easy to specify. Of the many different types of land-based pollution, one observer has offered the following classes based on parameters established by GESAMP.[41] They are (1) halogenated hydrocarbons; (2) petroleum and its derivatives; (3) other organic chemicals; (4) nutrient chemicals; (5) inorganic chemicals; (6) suspended solids; (7) radioactive substances and (8) thermal waste.

Each of the five categories of marine pollution being examined in this chapter are, of their own right, worthy of a much greater and more extended treatment than will be offered here. As in all endeavors of this kind, what is left out is often as important as what is included.[42]

The impact and consequences to the marine environment of land-based sources of pollution are both already profound and far-reaching. Gross has investigated the impact of sewage as a major source of contamination and found widespread effects. The United Nations Environment Programme also has an active research program detailing its consequences (see below).

"Wastes discharged to streams and estuaries by industrial and waste-treatment plants eventually reach the ocean. In the ocean, wastes can cause problems, in some cases killing organisms and in others rendering them unfit for human consumption." When the treatment facilities are constructed, properly maintained and staffed "... discharges of sewage wastes have relatively little long-lasting effect on water quality and on pelagic [open ocean] organisms which can swim out of the waste-containing waters." Most of the sewage disposal systems are engineered so that the effluents are diluted and dispersed into the receiving waters. "If the systems are properly designed and working, waste concentrations near the discharge point are relatively low. Later movements and mixing of these wastes by currents dilute them further..." There are problems with this scenario, though, and one of the concerns are the long-term effects of the sewage disposal systems themselves. As Gross points out, "Long continued waste disposal operations can cause drastic changes in the condition of the bottom and of benthic organisms which cannot move away from the waste materials. Among these are changes in the physical and chemical characteristics of the bottom..."[43] This brings into sharp focus a paradox contained within the Western industrial revolution paradigm. That paradox is this—the rate and extent of polluting human intrusions into the marine environment and global ocean have been increasing radically during the 20th century, and particularly in the last 45 or so years. At the same time, we are not aware of even many of the short-term effects of these contaminants (in large part due to a lack of global time-series data bases), to say nothing of what the situation will be like in the medium to long-term. As so many have remarked before, the future is being mortgaged without knowledge of what the payment schedule will resemble.

The problems associated with land-based sources of pollution are further complicated when particulates and waste solids are considered. Gross has concluded

> Many waste materials entering the ocean are either associated with particles or quickly become attached to solids. Waste solids usually settle out near the discharge points. Thus nearby waste deposits frequently contain high concentrations of carbon, bacteria and viruses as well as oils and waste chemicals. When the circulation near a contaminated bottom is sluggish, the consumption of dissolved oxygen by these deposits can cause depletion of dissolved oxygen in overlying waters... When the dissolved oxygen is used up, benthic organisms [die] ...[44]

The conclusion to this dismal picture is that the effect in coastal waters is to threaten the animals that live there, including ones habitually used for human food intake (fish, lobsters, clams and mussels, among others).

The problem is even more complicated, however. In most areas of the planet the conditions for the designed "safe" systems do not exist. Thus, the United Nations Environment Programme (UNEP), utilizing the expertise in its regional seas offices (see the following chapter below on IGOs where the UNEP will be more extensively discussed), found that each of the currently operating 10 regional seas programmes have all found sewage disposal to be a major problem and source of contamination.[45]

In 1982 GESAMP released a report entitled *The Health of the Oceans*.[46] This comprehensive study examined a number of important issues, including

> public health risks from micro-organisms ... [which derive in part from sewage treatment plants]. The ... report ... sounded the alarm on coastal pollution. It pointed out that although the threat was not in itself global, the combined effect on many local forms of pollution could become serious first on a regional *and later on a global scale*.[47]

Other studies have already found the threat to be a global one, and these will be discussed below.

A new consequence of run-off pollution from land may be in the incidences of algal blooms. A recent report of the Executive Director of UNEP described the phenomenon in this way:

> Algal blooms have occurred throughout history and are a natural phenomenon in marine waters. Unusual occurrences of algal blooms have been attributed to a combination of many factors, especially to disturbances in the marine ecological balance caused by climatic factors... The term "red tide" is in general use to describe the discolouration of water caused by any bloom, even though the algae that cause the tides can turn waters brown, yellow, green and other shades as well ...[48]

Some of these blooms are toxic, although most are not. What is so destructive to the marine environment, however, is the rate at which these blooms consume oxygen in the waters. Especially at the point at which the bloom begins to decay, the oxygen requirements go up exponentially and wreck the ecological balance of the area. Eutrophication occurs.

Many areas of the industrial world have seen the outbreak of such phenomena. In the summer of 1989 algal blooms in the Adriatic Sea were responsible for a loss of tourism revenues estimated to exceed $800 million.[49] The UNEP report referred to above noted many scientists involved in studying this problem feel the "... blooms may be nourished by pollution ... [and this position] is now gaining [favor] ..."[50] In the summer of 1989 the bloom stretched for over 400 miles in the Adriatic.[51] And such phenomena have not been limited to the Mediterranean area. Other blooms have occurred in 1987 along the coast of North Carolina, at Long Island in New York and in Guatemala (where 26 people were killed) due to paralytic shellfish poisoning.[52]

The consequences of algal blooms do not just reside in economic losses to tourism. As mentioned above, people can become ill and die. In addition, the loss to fish and other marine populations can be disastrous. In the Adriatic case of 1989 tons of dead clams and mussels washed up on local beaches.[53]

The phenomena of such "blooming" are not well understood by specialists. Although some experts have noted that "the rapid growth and decline of algae in the Adriatic is a natural phenomenon that comes in cycles, [the same experts note the 1989 occurrence] ... is only the third time in a century it has become a serious problem and never ... of this magnitude."[54] It is believed high levels of nitrates and phosphates (coming from agricultural fertilizers and household soaps) may have overloaded the local ecology. Honsell has stated: "The sea has passed its limits... The system cannot sustain such algae production... [The slime produced by the blooming] seems to produce a new ecosystem... [W]e have found unknown kinds of algae inside... [the mass of sea slime and gelatinous muck].[55]

Transboundary Transportation of Toxics

This category of pollution encompasses a wide variety of environmental threat to the ocean. In this section, however, I will not examine the important issue of the transportation of nuclear waste. That will be addressed in a separate discussion below. This part of the chapter focuses primarily on the movement of toxics from one nation to another or to the open sea. If a substance is toxic it is potentially poisonous, and an extreme danger to human health and other forms of life.

The problem of what to do with these substances has assumed enormous proportions. It is receiving widespread attention within various media and has become one of the primary concerns of numerous United Nations agencies, INGOs, developing and developed nations (although for different reasons). The concern over the existence and movement of toxics are chiefly related to toxicity levels created by chemicals and/or chemical processes.

The United Nations Environment Programme (UNEP) has estimated that there are more than 80,000 chemicals in every-day use throughout the planet.[56] These include industrial as well as non-industrial applications (the latter such as private home use and public recreation areas). The production and utilization of these chemicals in a wide variety of activities indigenous to industrial and post-industrial society generates very large amounts of toxic waste. Very little overall is being done to reduce the threat. Some basic issues surrounding this development are quite staggering.

> More than seven million chemicals are now known, and thousands of new ones are discovered every year.
>
> Many of these chemicals are potentially hazardous; some are in common use precisely because they are extremely toxic.
>
> Since the 1950s, chemical accidents have become more frequent and more serious.
>
> Stricter controls ... [in developed nations] encourage manufacturers to try and dispose of their wastes in developing countries ...[57]

While the figure of 7 million encompasses all known chemicals mentioned even only once in the literature, the much reduced figure of 80,000 chemicals in every-day use still presents an increasing problem to national governments in identification of substances in differing combinations, their regulation and disposal. These issues will be discussed below.

Many of these chemicals are deliberately released onto the land, and thereby enter the hydrological system. Atmospheric interaction at the point of land deposit also reappears as ocean pollution in particulate form. UNEP estimates that in the United States alone every year more than $13 billion "... worth of pesticides are used to eliminate weeds and pests that are in some way detrimental to human society ... [and notes] all pesticides, by definition, are toxic substances."[58]

> Not only has the number of chemicals increased dramatically over the past 20 years or so, but so have the quantities of them that are produced. Global production of organic chemicals, for example, increased from about 1 million tonnes a year in the 1930s to 7 million in 1950, 63 million in 1970 and about 250 million in 1985. Annual production now tends to double every seven or eight years.

This is being done with very little knowledge as to definitive long-term effects; whether these are of small doses accumulated over many decades or of repeated exposures, but at differing time variables and occurring throughout a subject's life. And these efforts continue whether that subject is human or another life form.[59] The latter point is relevant when considering concentration levels of chemicals that are found in fish and other sea creatures and when also consumed by humans.

The number of accidents involving chemicals have increased dramatically since the 1950s.[60] A few examples suffice. In 1959 mercury was discharged into waterways in Minamata, a city in Nigata prefecture of Japan, and 400 people died with approximately 2,000 injured. In 1973 in Fort Wayne, Indiana the compound vinyl chloride was implicated in the decision to evacuate 4,500 people because of a rail accident near population centers. The following year in the United Kingdom an explosion involving cyclohexane resulted in 23 deaths, 104 injuries and 3,000 evacuations. In 1978 in Spain, Mexico and Italy 316 people were killed, 350 injured and over 10,000 evacuated due to accidents involving propylene, gas and ammonia. In 1986 in Switzerland (Basel), a fire at a pesticide plant resulted in the release into the Rhine River of large quantities of toxic chemicals. More incidents occur almost every week in every region of the globe.[61]

Among other environmental groups, the international non-governmental organization (INGO) Greenpeace has conducted extensive monitoring efforts on the international trade in toxics. Developed nations, often because they are prohibited by factors of cost or national legislation in their respective countries, export to developing nations the dangerous wastes generated by their own industrial processes. In 1988 Greenpeace prepared an extensively documented report for *The Ad Hoc Working Group of Legal and Technical Experts with a Mandate to Prepare a Global Convention on Control of Transboundary Movements of Hazardous Wastes*, convened by the United Nations Environment Programme in Geneva. It was noted in that report that almost 3,700,000 tons of wastes were transported from industrialized to non-industrialized nations between 1986 and 1988 alone.[62]

The issue is particularly sensitive to nations in Africa, although no continent can escape the consequences of unregulated short-sighted economic and industrial processes. The Economic Community of West African States (ECOWAS) met in Togo, during the summer of 1988, ostensibly to promote economic solidarity in the seemingly never-ending rounds of trade negotiations between developed and developing nations. The trade in hazardous waste, however, emerged as a major issue. At this meeting, the ECOWAS representatives adopted a resolution criticizing this export trade. "A statement by a member country represented the ECOWAS position: 'We cannot accept that at a time when industrialized nations refuse to buy our commodities at reasonable prices, these same countries are selling us death for ourselves and our children.'"[63] This is an emotional, as well as a scientific, environmental and public health issue. It will not go away. Developed nations continue to seek out new locations for the deposit of the dangerous by-products of industrial and post-industrial society, while developing nations are constrained by low standards of living. These poor nations need the money, hence contracts are offered and letters of agreement signed.

Carroll has written on the definitional components to the transboundary transportation of environmental problems. His definition is incomplete, because it does not take into account the reality of even unperceived environmental problems. In other words, because individuals do not "perceive ... [or] believe" a transboundary pollution problem does not exist does not mean there is no transboundary pollution problem.[64] The author does qualify his definition with the word "generally," but this caveat cannot escape the difficulty. There is much to be said for his statement that in politics "perception thus becomes reality." Often this is quite true. But in a political-scientific nexus the reality of science (and here I refer to chemical dispersion over or through national borders, as but one example) exists independently of the recognition that a problem may now be political. Whether individuals or governments or other organizations are aware or not that a certain transboundary transportation problem exists is, in this sense, immaterial. The problem can be present without the recognition or exact identification as to the cause(s). Consequences of environmental policies may not become manifest for years or decades later, and here it is correct to speak of perception or belief affecting political action. But to state that "... if transborder damage goes unrecognized by the populace affected, then there is generally no transboundary environmental problem" is wrong because knowledge is a reality perdurable whether or not humankind at a certain point or stage in its intellectual development is aware of it or even recognizes it through existing scientific means. It is quite credible to argue that human beings are deleteriously affected by damaging environmental substances (read: possible transboundary transportation of toxic materials) this very day, now, of which we know nothing or next to nothing about; yet their identification may be only a matter of years or decades away. Indeed, a constant attention to the press and other media, and continual reading of relevant journals, points up incidents too numerous to cite that demonstrate what was thought to be safe and non-harmful to humans and other life forms now is not so—and this is due to the unearthing or discovery of new information.

Oil

Pollution from the transportation of oil in ships, spills or accidents (also involving shipping), discharges from the normal operation of drilling platforms and, to a lesser extent, the deposition of tar balls on beaches and coastal areas (primarily from shipping and drilling activities) are all increasingly controversial economic consequences of an industrial world dependent on fossil fuel for energy.

Spectacular accidents involving tankers have generated wide-spread and negative publicity for oil companies and national governments responsible for regulation of safety issues and environmental threats to coastal areas. R. B. Clark has identified four dramatic incidents of this type of oil pollution, to which can be added a fifth.

> Oil pollution of the sea attracts great public attention because it is visible and most people encounter it, either at first hand on bathing beaches or ... from pictures on television or in the press...
>
> In March 1967, the tanker *Torrey Canyon* carrying 117,000 tonnes of Kuwait crude oil was wrecked off Land's End, depositing 40,000 tonnes of oil on the beaches of Cornwall and Brittany.
>
> In April 1977, an offshore well in the North Sea Ekofish field blew out and in the seven days before it could be brought under control discharged an estimated 20,000 - 30,000 tonnes of oil into the sea.
>
> In March 1978, the *Amoco Cadiz*, carrying 223,000 tonnes of Arabian and Kuwait crude oil was wrecked on the coast of Brittany, causing enormous pollution of beaches there.
>
> In June 1979, the *Ixtoc I* oil well in Campeche Bay off the Mexican coast blew out and could not be completely sealed until March 1980. During those nine months it discharged an estimated 350,000 tonnes of oil into the Gulf of Mexico.[65]

And of course, in March 1989 the *Exxon Valdez* ran aground in Prince William Sound, Alaska and discharged almost 11 million tons of oil on pristine beaches and breeding grounds for fish and birds and other wildlife.[66]

The intense political controversies generated by these accidents, so far usually confined to national, i.e. domestic—as opposed to regional or global—political debate, raises an interesting question. Would repeated or frequent accidents raise regional or global political concerns? Certainly the role of the planet's telecommunications infrastructure disseminates almost instantaneously the news of such events and visual documentation of devastating "localized" impacts. The strictly scientific potential for such a "globalization" would depend on the frequency, duration and location of the spills. Such a development is unlikely. A complete break-down of safety procedures and engineering designs would be necessary before such an eventuality could take place. However, political concern over the environmental effects of such spills or accidents in the industrial market economies could very easily grow and build-up to overcome the heretofore minuscule multilateral "spill-over" of such ecological disasters.

As in all types of pollution affecting the ocean, even when a relatively short to medium term problem is not felt to exist, the possible problems arising to the long-term situation from constant low-level indicators are another matter entirely. In 1976 Edward Goldberg recognized this problem in one of the early studies to address the global ocean as a total planet-wide system.[67] Although his efforts inevitably reflected the state of knowledge then present (a situation somewhat analogous to today, but sufficiently modified to allow for a substantial increase in our understanding of planet-wide oceanographic processes), *The Health of the Oceans* is an important methodological statement of the norms and procedures necessary for an understanding of the threats present to human and other forms of life which ocean pollution threatens.

The chapter on petroleum hydrocarbons is a case in point (Goldberg, Chapter 6). The author was greatly constrained at the time in his work by the state of satellite technology and the levels of computer sophistication present in both software and hardware. Global study programs monitored or administered by international agencies did not possess the scope of involvement or sheer size and number which characterized the international multi-agency efforts of the 1980s. This issue will be taken up in the last section of this chapter. Nevertheless, Goldberg was able to find that there were many threatening dimensions to the problem—"[m]arine organisms now carry body burdens of petroleum hydrocarbons disseminated by ... [humans]."[68] He was very concerned by this, as well as other issues; but was also aware of the analytical problems present when attempting to address an issue as complex as marine ocean pollution. Such problems plague the specialist today.

> An assessment of the seriousness of the hydrocarbon pollution problem is made difficult in a number of ways. First of all there are the analytical problems. Crude oils are composed of many tens of thousands of compounds and vary in composition according to their source... [Second, there] ... are three general sources of petroleum-containing hydrocarbons to the ocean system: ... [human] generated substances; hydrocarbons produced by the marine organisms; and those ... that seep in naturally through the sea floor. The origins of any one sample can be difficult to determine.[69]

Some of these issues have been ameliorated since Goldberg wrote his study in the middle 1970s, and will also be addressed in the following chapter on selected United Nations and affiliated agencies.

An additional problem in the hydrocarbon issue, as well as other forms of pollution, relates to the partial dichotomy between coastal and open ocean bodies. Goldberg sees the question this way: "The open ocean differs significantly from its coastal counterpart not only in its time scales but also in its relationships with the continents and sediments. The time spans describing natural processes extend from hundreds to hundreds of millions of years as contrasted to months to decades in the coastal ocean." Aluminum, for example, flows into the ocean from "weathering processes" present on continents. These particles exist for perhaps 100 years as a solution before descending into sedimentary layers. Sodium, on the other hand, a major component of sea water, exists for approximately 100 million years in the open ocean before transferring into sediment.[70]

Significant attention has also been focused on the problem of tar pollution, particularly in the Mediterranean. The United Nations Environment Programme describes the problem this way in the summary of a recent report: "Tar balls in the marine environment are a derivate of oil or oily compounds which were released into seawater, lost their volatile fraction by evaporation and turned into soft black lumps called tar balls. Due to the relatively high intensity of oil activity in the Mediterranean Sea, tar pollution became a problem... The severity of this problem is accentuated by the fact that tar on the beach is a serious nuisance to sea bathers, and the Mediterranean Sea is becoming more and more attractive to coastal oriented tourism."[71] It is not known, however, whether the problem is beyond that of aesthetics. Golik found that

> Very little is known about the effect of tar on marine organisms. In a study aimed at finding whether tar constitutes a threat to marine life, *Zsolnay et al.* (1978) examined concentrations of tar and of aromatic hydrocarbon, which is the most toxic component in crude oil, at the same stations. They found that there is no relationship between the two, presumably due to different dispersion mechanisms.[72]

The sources of tar bar pollution are diverse. Golik's findings agree with those of Goldberg that the tar can come from natural seeps in the sea floor, from oil tanker accidents, intentional release from ships and from land-based sources.[73]

As in all scientific studies, methodology is a primary concern. Golik identified a significant problem with some previous studies of the amount of tar present on beaches. "One should distinguish between the rate at which tar lands on the beach and the standing stock of tar on the beach."[74] Tar landing on a beach is a function of the "dynamics" of ocean or sea circulation. However, tar can already exist on a beach previous to a study attempting to quantify the problem. "... [O]ne must realize that waves cause lateral movement of tar balls along ... [a] beach. Therefore, repeated sampling of a narrow beach strip yields the rate of tar accumulation from both the sea and the beach on either side of the sampling strip." It has been demonstrated "... that even under conditions of ... [a] calm sea, tar balls moved laterally by the small waves [present] along the beach..." Relatively simple matters such as this can haunt specialists engaged in pollution research.

Golik's study is one of the few—because of its narrow and specifically defined focus—that can answer the question "Has there been any reduction of tar pollution in the Mediterranean?"[75] Evidence appears to indicate a tentatively affirmative answer. The reasons are multifaceted. First, the second oil crisis of 1978/79 (to be distinguished from the first following the Arab-Israeli war of 1973) resulted in a significant increase in petroleum prices. Specifically, within the Western industrialized market economies this contributed to an overall reduction in oil consumption, a significant increase in conservation efforts towards greater energy efficiency (particularly within the United States) and a "reduction in marine transportation ..." of petroleum. All this "... brought about a series of activities, technological and legal, which caused a reduction of oil released into the sea and increased the awareness of marine pollution [witness certain amendments to international treaties, specifically OILPOL 54]". Utilizing a British Petroleum statistical review on world energy, issued in 1985, but also noting that the indicators and actual numbers of other studies are very few, Golik concludes that while some signs are "encouraging" it is still too early to state any definitive answers. He calls for more studies in other areas of the Mediterranean where tar ball observation programs have not been instituted.[76]

Nuclear Waste

The disposal of the by-products created in the nuclear fuel cycle is a serious and long-standing problem which has affected the world community since the dawn of the Nuclear Age. The problem has grown significantly in the last several decades. There are three general levels of waste as defined by the International Atomic

Energy Agency: high-level radioactive waste (HLW), intermediate-level radioactive waste (ILW) and low-level radioactive waste (LLW).[77] Calmet has noted that the first dumping of LLW occurred in the North East Pacific Ocean in 1946.[78]

Social scientists have, in recent years, become more and more interested in the political problems associated with the disposal or dumping or radioactive materials. Nurmi has noted, for instance, that

> The OECD Nuclear Energy Agency has supervised such dumping in accordance with the requirements laid down by the IAEA in the Multilateral Consultation and Surveillance Mechanism adopted by the OECD Council in 1977. According to OECD estimates, between 1967 and 1982 the tonnage dumped annually remained roughly the same while the radioactivity level of waste increased. The dumping ... was temporarily halted in 1983 and 1984 due to recommendations adopted by the Consultative Meeting of the London Dumping Convention. In September 1985 the permissibility of dumping of low and medium level radioactive waste was considered again.[79]

At a subsequent meeting of the London Dumping Convention (the Ninth Consultative Meeting), the parties to the agreement decided disposal of nuclear wastes on land was a "... safer and more controllable ..." process than disposing of the material in the ocean. However, not all nations held to this position. Marine disposal of nuclear waste is a continuing policy of many states. Since the focus of this research is on the degradation of the ocean environment primary emphasis will remain there and not with land disposal alternatives (which is the other alternative to disposal at sea).[80]

The saliency of the disposal at sea issue is aptly illustrated by a decision of the United Kingdom government in the late winter of 1990 to refrain from signing a final communique drawn up by environment ministers meeting at The Hague to declare that "the North Sea is not suitable for the dumping of radioactive waste nor for disposal of such waste into the seabed."[81] The British environment minister noted his country does not dispose of nuclear waste in the sea, but was committed to the principle of keeping the option open. The communique spoke of agreement or consensus within "competent international organizations" that disposal of the by-products into the North Sea specifically was "not suitable." The British delegation insisted on appending to the document a statement that "The United Kingdom was unable to accept this provision on [the] ground that competent international authorities have not declared this method of disposal unsuitable." This issue of whether or not the scientific community, of which scientific IGOs are an important part, has developed broad agreement on the question of nuclear dumping at sea is another critical component of the debate now confronting the industrial world.

In 1990 GESAMP (Joint Group of Experts on the Scientific Aspects of Marine Pollution) issued a comprehensive report on the state of the marine environment.[82] That research noted the often overlooked fact that "[r]adioactive substances are present naturally in ocean waters."[83] The amounts of radioactive substances and radionuclides in the atmosphere and ocean are, however, increased by the actions of nation-states. The categories of fall-out from nuclear weapons tests, effluents from nuclear power and reprocessing plants, and dumping of LLW at sea are cases in point. Although it is important to note radioactivity is present in the oceans as a natural occurrence, it is also important to measure the increases (as accurately as possible) caused by human intervention.

Nuclear weapons tests were the primary cause for the increase in artificial radionuclides introduced into the marine environment after 1945. The Nuclear Test Ban Treaty, signed in 1963, was the major reason for the subsequent reduction; although not all the major nuclear powers signed (France being the main exception). Ironically enough, the GESAMP report notes "[t]he inputs to the oceans in the southern hemisphere have been half of those in the northern hemisphere.[84] Of course, most of the nuclear power states are in the north. This does not mitigate the irony, however, that in the quest to achieve higher and more secure levels of perceived national security the industrialized and developing nations have significantly contributed to the degradation of the ocean nearest their own borders.

Additionally a higher concentration in one half of the planet does not, of course, mean that dispersal patterns have not affected other regions as well. The report noted that an International Atomic Energy Agency (IAEA) Expert Group conducted experiments to measure concentrations of radionuclides in the ocean. The results obtained showed

> ... slightly higher estimates of the ocean inventories of these radionuclides on the basis of measured concentrations. Fall-out from nuclear tests has been the only source of world-wide radioactive contamination of the oceans. However, its widespread dispersal has resulted in low-level ambient concentrations and hence in negligible additions to the exposure from natural background.[85]

It should noted, though, that although the report states *overall* exposure is "negligible" world-wide, this is an average and it is not clear what the medical effects to human or other life forms are of long-term but low-level exposure dispersed in this manner. There are other impacts on the marine environment besides nuclear weapons tests, which have now moved underground in the United States and the Soviet Union, since the ratification of the test ban treaty.[86]

The thermal discharges from operating nuclear power plants "... make an additional contribution to the radionuclide inventory;" although presently available information makes the impact site-specific rather than one larger in scope, i.e. regional or world-wide.[87] These reports, it should be noted, almost always contain an important disclaimer. With specific reference to the effects of thermal discharges, it is stated that

> A large body of information is available in the literature concerning the effects of thermal discharges. Many studies refer to the *overall consequences* of station operation. To understand these effects and their causes it will be necessary to distinguish the separate mechanical processes of cooling water use and cooling water treatment. A major difficulty in interpretation is that the *extensive background information needed to provide perspective and to evaluate changes is seldom available*, even in regions of long-standing technological activity.[88]

In other words, the findings of these bodies are tainted to the extent they do not provide long-term and detailed information over a wide spectrum or issue area of related scientific concerns and questions. Finally, the most serious component to the question of nuclear waste and the ocean relates to the dumping at sea of radioactive refuse. Of all the potentially fateful steps humankind has made regarding the ocean environment, this one is the most curious and uncertain.

As I have indicated, there are many components to the "pollution problem" which inhere in the industrial paradigm now present in the western world. Calmet states no *high-level* radioactive waste has been dumped into the ocean but, as was noted earlier, low-level amounts have been introduced beginning in 1962 and ending in 1982.[89] This is not the place to enter into a discourse on the specific scientific principles regarding levels of radioactive waste and whether those levels are, given current knowledge, "safe." For instance, Calmert mentions in her article that between the two above dates LLW at "... an estimated 63 BPq (1.7 MCi) of radioactive waste coming from research, medicine and nuclear industry activities ... [were] packaged, usually in metal drums lined with ... concrete and bitumen matrix, and disposed of at sea."[90] Is this sufficient for safety purposes? The judgment is a controversial one, particularly regarding the impact on marine food resources.

Zhu and Chan have also provided comparative tables for projected waste volumes for nuclear power plants, as one component of the nuclear process cycle, using thousands of cubic meters as a base measurement and comparing the projected years of 1988, 1990, 1995 and 2005. Four boxes are presented—projected LLW, projected ILW, projected HLW and projected spent fuel volumes. The Organization for Economic Cooperation and Development (OECD) countries in Europe, the Pacific and North America are represented by bar graphs. All the projected levels for waste and spent fuel categories increase significantly from 1988 to 2005, including the OECD nations in North America. The latter also show an increase, although much more modest.

What immediately concerns some specialists most profoundly in this controversial issue are the questions of leakage and the effects this would generate on the food chain. The GESAMP report on the state of the marine environment (1990) succinctly states the problem.

> Doses to, and effects on, marine organisms or marine populations are much less well known [than the effects on human beings]. As in ... [humans], effects may be somatic (in the individuals exposed) or genetic (in the germ cells of the irradiated individuals and therefore transmissible to their descendants). While for ... [homo sapiens sapiens] the individual is the target of concern, for marine organisms interest is primarily in population effects such as survival, growth and reproductive performance. However, at the levels which the populations of marine organisms are currently exposed as a result of ... inputs of radioactive materials, adverse effects have not been observed.[91]

One of the points of contention in this debate concerns just what is the appropriate level of adversity that should be identified as it relates to marine organisms. Another concern relates to the presuppositions of the original disposal-dispersal schemes which viewed the ocean as "... a large dilution sink ... [which would ensure] low concentration rates [of radionuclides] in the marine environment."[92] The current state of knowledge now accepts that the ocean is not a "disposal unit." It is not something that can be utilized as a "drain-pipe" to take away to some far-away place the unwanted and dangerous by-products of post-industrial societies.

The Atmosphere

A major source of degradation of the ocean are pollutants carried in the atmosphere. These source elements are generally, although not exclusively, created by the industrial and agricultural processes extant on the planet today and originating in the Industrial Revolution of the previous century. Naturally produced environmental changes which impact on the ocean are of more long-standing and less catastrophic duration insofar as their intensity to affect human life is concerned. Examples of the latter category would include volcanic eruptions, forest fires or soil run-off from rivers, lakes and streams into the global ocean. As the recent GESAMP report makes clear

> The atmosphere carries material from many sources, natural and anthropogenic. The former include dusts from arid areas, soil, volcanoes, vegetation and forest fires, as well as aerosols from the oceans. Among the anthropogenic sources are emissions from industries, energy production and use, waste combustion and agricultural activities. Sources of synthetic organohalogen compounds to the atmosphere include emissions during production and disposal, as well as during applications, e.g. of pesticides.[93]

A 1980 GESAMP study, with primary support from the World Meteorological Organization, studied in depth the interchange between oceanic and atmospheric variables as they effect the state of the ocean. This air-sea interchange "... and the identification of those pollutants for which an atmospheric pathway might be important in causing adverse effects on marine resources and amenities..." is an area of critical research.[94] "The most important mechanisms for the sea-to-air transport of *nongaseous pollutants* is the air bubble, which scavenges interfacially active material during its rise through the water and ejects some of the absorbed material into the atmosphere when it bursts at the sea surface."[95] Research focuses on the exchange mechanisms present during this process. In addition, there are gaseous pollutants which exchange at the interface between the ocean and the atmosphere. As the report noted, "... [t]he magnitude of oceanic bubble transport for gases is unknown but thought to be less than that due to molecular exchange."[96] It is beyond the scope of this chapter to enter into a detailed examination of the physical and chemical properties which do interface between the various metals, sulfates, radionuclides, micro-organisms or gaseous hydrocrabons or halogenated hydrocarbons (to cite just a few examples) that reside in the atmosphere or ocean. However, it is primarily the spin-offs from industrial activities, and not the naturally occurring physical and chemical processes of the ocean, that are of concern.

One of the areas in which very little knowledge is known is the effect on the ocean of the approximate century-long introduction of high levels of carbon dioxide into the atmosphere. A recently published brochure by the International Geosphere-Biosphere Programme (IGBP), which is coordinated by the International Council of Scientific Unions (ICSU), poses the very apropos question: "How do ocean biogeochemical processes influence and respond to climate change?[97] In other words, is a change occurring in the world's climate which could significantly influence the ocean? There are many problems facing researchers in their attempts to answer this question. One of the most difficult was stated by the GESAMP environment report of 1990.

> The total atmospheric input of chemicals to the ocean surface is the sum of the amounts entering directly in gas and particulate phases (dry deposition)

> and those falling out as rain and snow (wet deposition). It is difficult to obtain an accurate estimate of either. Even when contaminant-free samples of precipitation are available, and this is particularly difficult to achieve over the oceans, interpretations of the data is complicated because chemical composition varies with differences in vertical distribution, duration, intensity and droplet size of the precipitation.[98]

Climate change or global warming—as one important part of the world-wide change in ecosystem which human beings have only of late introduced—is certainly the most serious threat to Planet Earth since the Industrial Revolution spread its radically efficient form of economic organization. Indeed, such a note is a gross understatement. The Industrial Revolution occurred only 200 years ago. The biosphere, of which human beings are of course an integral part, has never seen anything like global warming *in such a short period of time*. On the geologic time-scale, two hundred years is but a second.

The National Research Council in 1988 issued an initial report on possible contributions of the United States Government to the IGBP.[99] Noting the impact of carbon dioxide release into the planet's atmosphere, the report's summary of recommendations contained the following warning:

> Human activities are currently leading to changes in the global environment at virtually unprecedented rates, with potentially severe consequences for our future welfare. Increases in carbon dioxide and other greenhouse gases, concerns for climate change, the appearance of the antarctic ozone hole and worldwide depletion of the ozone shield, ... and a host of other changes in our environment have captured the attention of ... [many]. *The problem of global environmental change is crucial and urgent.*

There is an interactive process present between the ocean and climate. Changes in the atmosphere directly impact on the ocean environment, and there is a corresponding relationship between the hydrosphere (of which the ocean is the overwhelming component) and the atmosphere. The problem for scientists and analysts is that so little is known of the interchange, save that it exists. As Dickinson has observed, "[c]urrent information on the global and regional budgets of water is inadequate, and key ingredients—i.e., clouds, atmospheric water vapor, soil moisture, precipitation and evaporation—are now inadequately measured. These variables and the processes responsible for them need to be better represented in global models."[100]

The IGBP and the International Council of Scientific Unions, among other programs and IGOs, will be discussed in the following chapter. However, for now it is sufficient to state some of the general concerns and questions international cooperative efforts like the IGBP are addressing in an attempt to understand the effect of ocean biogeochemical processes and the urgency of climate change research. The ICSU brochure states the pollution issue succinctly regarding the role of the ocean in the total Earth System.

> Through a combination of physical and biogeochemical processes the oceans function as both source and sink for carbon dioxide. Thus, the oceans play a key role in regulating the climate system, partly through biological processes such as marine primary production. Climate change will alter marine biogeochemical cycles [likewise, degradation of the ocean environment will alter systems of marine primary production], which in turn will affect the physical aspects of climate to an extent that is as yet unknown. Human-induced changes on land will influence coastal biogeochemical cycles, and expected sea-level rise, and expected sea-level rise, as a consequence of climate change, will have profound effects on coastal ecosystems.[101]

The impact of the projected global warming for human life is difficult to exaggerate. It is also difficult to estimate accurately. Sea-level rises will threaten to inundate coastal areas and cities on an unprecedented scale; this much is absolutely certain if projections are even remotely accurate. Large metropolitan areas in the industrial market economies of the West will face disaster.

Climate and pollution are inextricably intertwined. It is more than fairly certain climate changes are already occurring over the planet which threaten to upset carefully balanced regulative life-support systems.[102]

Recent work has been conducted on expected implications of climate change in the Mediterranean region. This is a useful area in which to focus because long-term, cautious and thoroughly documented studies have already been conducted under the auspices of the United Nations Environment Programme for a number of years. The expertise is already in place and has been working efficaciously for over a decade. A number of tentative conclusions regarding the global warming *problematique* and the consequent ocean degradation can be stated. The preface to a 1989 report noted

> In spite of uncertainties surrounding the predicted climate changes, greenhouse gases seem to have accumulated in the atmosphere to such a level that the changes may have started already and their continuation may now be inevitable... The environmental problems associated with the potential impact of expected climate changes may prove to be among the major environmental problems facing the marine environment and adjacent coastal areas in the near future.[103]

Sestini, et al. present an alarming picture of climate-induced environmental changes in the Mediterranean region. Such alterations will subvert social, economic and political relationships not only within countries but between them as well; and undoubtedly will move outside the regional context quite quickly and elevate themselves to the international realm. The present state of the environment in this area has become "extremely precarious" as a result of the constant interference of natural processes.[104] The authors note "[o]ver the centuries there has been an increasingly serious interference with natural balances throughout the Mediterranean basin. This has altered the original aspects of the countryside and brought about, through a constant process of deterioration, extremely precarious situations requiring urgent and appropriate remedies." The situation fares no better on the coastal areas.

> Deltaic and lowland coasts are presently experiencing serious environmental problems, due not to recent climatic oscillations, but to careless development of land use over the last 50 years. These include the urbanization of the coast, the building of deep harbours and coastal defence structures, a decrease of lagoon surface (via land reclamation), ... the deterioration of rivers' water and solid discharges, pollution of coastal waters, ... etc.

The primary source for pollution in the Mediterranean is the run-off from land-based sources.[105] This includes "... domestic sewage, industrial discharges and agricultural run-off."[106] Here, deterioration of the ocean environment is consistent with other global trends.

Sestini et al. document a large number of changes in the Mediterranean basin likely to result from the greenhouse effect. They are too detailed to discuss here, but a few examples will illustrate the point (by looking at temperature, impacts on ecosystems and impacts on agriculture and fisheries).

Carbon dioxide concentrations in the global atmosphere have increased more than 32% between 1900 and 1985.[107] Most of this is attributed to nitrous oxide and chlorofluorocarbons emitting from industrial enterprises. Average global temperatures have been steadily increasing since the late 19th century. "Overall there has been a warming of about 0.5 degrees C in the past 100 years." The authors go on to declare that there exists within the scientific community a consensus that "... if allowed to continue to build up, a doubling of the greenhouse gases concentration (relative to the pre-industrial era) will occur sometime in the 21st century, possibly as early as 2030 AD." Furthermore, "[a] corresponding global increase of temperature of between 1.5 degrees - 4 degrees C is predicted, to become effective 2-3 decades later, in consideration of the lag in homogenization effect."[108] What effect will this likely have on ecosystems in the basin area? In all probability the impact will be significant.

> Shallow onshore marine areas would become warmer, and more saline, locally hypersaline. In the ocean a 2 degree C rise in air temperature would result in a water temperature rise of 0.8 degree C–1.5 degree down to 200 m between Lat. 30 degrees N and 30 degrees S; [and even] surface currents would change over large parts of the oceans, possibly also in the Mediterranean.[109]

This would change the survival parameters of many fish and other aquatic creatures. Sestini et al. note that "[i]n some areas even a 1 degree C rise might have a marked adverse effect on fish life because of changes of oxygen concentration and changed water chemistry."[110]

Changes in weather patterns will directly relate to productivity and the viability of several agricultural crops. In the Mediterranean this concerns principally the following: "... wheat, soya beans, sugar beets, tomatoes, tobacco, citrus and other fruit trees."[111] Predictions of increased salinity levels in soils are also expected to have negative effects because of the availability of "... fewer nutrients being available to ... plants."[112]

The Mediterranean is a large region, complex and not just confined to what some people would think of as economically backward or underdeveloped southern nations. The area includes North Africa and the Eastern Mediterranean, including Greece; but it also includes Northern Italy (an area of high industrial production, and concomitantly high levels of pollution as well) and Southern France. But all of these nations' coastal areas influence the surrounding regions and areas far beyond. For example, Northern Italy influences, and is influenced by, the ecological variables that occur within Switzerland, Germany, Yugoslavia, Albania and other states in Eastern Europe. The "first law of ecology" is valid here as well as everywhere else on the planet: Everything influences everything else. Put another way, all things relates to each other. Humankind, through its everyday actions, has yet to internalize through its social processes this physical law ... it avoids doing so at its continuing peril.

Chapter 3 Notes

[1]This section on oceanography draws primarily from the substantial work of M. Grant Gross, *Oceanography, A View of the Earth*, 4th ed. (Englewood Cliffs, NJ: Prentice-Hall, Inc., 1987).

[2]Ibid., p. 32.

[3]See Barbara Ward and Rene Dubos, *Only One Earth* (New York: W.W. Norton & Co., Inc., 1972), pp.195-196.

[4]Harold and Margaret Sprout, "Environmental Politics: What Role for Political Scientists," in Stuart S. Nagel, ed., *Environmental Politics* (New York and London: Praeger Publishers, 1974).

[5]Andrew Maguire and Janet Welsh Brown, eds., *Bordering on Trouble: Resources and Politics in Latin America* (Bethesda, MD: Adler and Adler, Inc., 1986).

[6]Gross, op. cit., pp.187 and 176 where, on the latter page, the following analysis on currents is provided: "Currents—large scale water movements—occur everywhere in the ocean. Surface currents are driven primarily by winds. Subsurfacc currents, on the other hand, are driven by the sinking of chilled waters in the subpolar oceans, which then spread out to flow through all the oceans and eventually return to the surface to be warmed. Both surface and subsurface currents are caused by the unequal heating of the earth's surface. Ocean currents transport heat from the tropics toward the poles, thereby partially equalizing surface temperatures over the earth. Ocean currents play an important but little understood role in controlling climate..." Some authors follow the practice of speaking of several "oceans" instead of one Global Ocean. For reasons made clear in the body of this chapter, I adopt the latter description.

[7]They are made up of representatives from the following sponsoring agencies: International Maritime Organization (IMO), Food and Agriculture Organization (FAO), United Nations Educational, Scientific and Cultural Organization (UNESCO), World Meteorological Organization (WMO), World Health Organization (WHO), International Atomic Energy Agency (IAEA), the United Nations (UN), and the United Nations Environment Programme (UNEP). See *GESAMP-IMO/FAO/ UNESCO/WMO/WHO/IAEA/UN/UNEP*. [Hereinafter referred to as GESAMP] *Joint Group of Experts on the Scientific Aspects of Marine Pollution. Report on the Eighteenth Session, Paris, 11-15 April 1988. Rep. Stud. GESAMP, No. 33, 1988.*

[8]Satu Nurmi, "Issues and problems in the protection of the marine environment," in John E. Carroll, ed., *International Environmental Diplomacy, The management and resolution of transfrontier environmental problems* (Cambridge: Cambridge University Press, 1988), p. 208.

[9]Gross, op. cit., p. 268. In addition, see Andrew Bakun, "Global Climate Change and Intensification of Coastal Ocean Upwelling," *Science*, 12 January 1990. Bakun notes in his Abstract: "A mechanism exists whereby global greenhouse warming could, by intensifying the alongshore wind stress on the ocean surface, lead to acceleration of coastal upwelling. Evidence from several different regions suggests that the major coastal upwelling systems of the world have been growing in upwelling intensity as greenhouse gases have accumulated in the earth's atmosphere. Thus the cool foggy summer conditions that typify the coastlands of northern California and other similar upwelling regions might, under

global warming, become even more pronounced. Effects of enhanced upwelling on the marine ecosystem are uncertain but potentially dramatic." Gross, op. cit., pp.298 (emphasis added) and 300 has the following to say regarding upwelling: "[A]reas of upwelling are characterized by a large proportion of phytoplankton (diatoms) that aggregate into clumps or long chains. They can be eaten directly by schooling fishes or by small herbivorous zooplankton... [H]ighly productive ocean areas commonly support short food chains of rapidly growing organisms where transfer efficiencies may be 20% or higher... *About half of the world's fish catch comes from upwelling areas.*" In addition, "Fish and other animals are distributed in the upwelling areas to take advantage of rich production of phytoplankton. Filter-feeding fish (sardines, anchovies) live in the areas of maximum algal production. Farther offshore other herbivores (copepods) feed on the algae and they, in turn, are eaten by carnivorous fish (mackerel). The largest zooplankton (euphausiids) live near the edges of continental shelves and are eaten by the largest fish (hake)." The examples noted in parenthesis are just that, examples of the point Gross is making.

[10]Gross, op. cit., p. 268.

[11]A November 1987 case of a Dutch cargo ship being held on bail by Egyptian authorities on the charge of polluting the Egyptian coastline with fuel oil is a case in point. The Egyptians wanted $97 million US in damages for the accident. See *UNEP Environmental News, February 1988 - Supplement.* Quoted from *Reuter*, 8 November 1987.

[12]Gross, op. cit., p. 269. Although this an extreme example the point is well made.

[13]J. Pernetta and G. Sestini, "The Maldives and the impact of expected climatic changes," *UNEP Regional Seas Reports and Studies, No. 104, UNEP, 1989*, p. 1. Also, James M. Broadus, "Impacts of Future Sea Level Rise," in Ruth S. DeFries and Thomas F. Malone, eds., *Global Change and Our Common Future, Papers from a Forum* (Washington, DC: National Academy Press, 1989).

[14]This paragraph is drawn from Gross, op. cit., pp.163-164.

[15]This development is reported in *New Scientist*, 10 February 1990, p. 34. See also *New Scientist*, 10 March 1990, p. 28 for a report on a new UNEP system which promises "to provide a reliable early warning system on the health of the planet, using a mammoth computer database it has been compiling for the past five years."

[16]Gross, op. cit., p. 163.

[17]Ibid.

[18]Gross, op. cit., pp.163-164. This outstanding oceanographer notes, "In the *pycnocline zone*, water density changes markedly with increasing depth and thus forms a layer of much greater stability than the overlying surface waters. Formation of a pycnocline zone results from marked changes with depth in either salinity or temperature. Marked temperature changes with depth are common in the open

ocean... There the pycnocline coincides with the *thermocline*, or zone of sharp temperature change. Thermoclines are especially prominent in open-ocean areas, where ... surface layers are strongly heated during much of the year."

It is useful for the social scientist dealing with these issues of ocean degradation and the role of the state system to understand some basic facts regarding the ocean realm he or she is ultimately trying to protect. Because of that, the following lengthy quote is also offered the interested reader. "At higher latitudes there is less surface heating by incoming solar radiation. There the lowered surface salinity caused by abundant precipitation causes of *halocline*, a marked change in salinity with depth. The halocline is well developed over much of the North Pacific Ocean ... where it is an important factor in the stability of the pycnocline zone. The halocline is also important in most coastal ocean areas.

"The pycnocline ... partially isolates the deep ocean from surface processes because low-density surface waters cannot readily move downward through it. Deep-ocean waters move slowly upward through the pycnocline as part of the worldwide deep-ocean circulation. This is a significant factor in oceanic circulation at mid altitudes and low altitudes. Seasonal changes in temperature and salinity are difficult to detect below the pycnocline."

"The *deep zone* contains about 80% of the ocean's waters and lies at depths of about 1 kilometer [.621 miles] or more—except at high latitudes, where deep waters are on contact with the atmosphere. The relationship between deep-ocean waters and high-latitude areas of exposure to the atmosphere accounts for the low temperatures (average 3.5 degrees C) [38.3 degrees F] of sub-surface ocean waters ..."

"Waters in the deep ocean occur in layers. Careful study of temperature and salinities permit different water masses to be identified. Each water mass is characterized by a temperature and salinity (or a narrow range), which permits tracing movements of waters from their sources."

[19]See Charles W. Kegley, Jr. and Eugene R. Wittkopf, *World Politics, Trend and Transformation* (New York: St. Martin's Press, 1981), p. 28. The authors note "high-politics" have traditionally centered around questions of military-strategic variables, while "low politics" have addressed international economic issues. One could add, however, that the category of "low politics" has also contained the contentious areas of environmental matters, immigration and technology transfers.

[20]Gross, op. cit., pp.357 and 384. Mariculture is a large industry in Japan. In the early years of the last decade there were over 30,000 enterprises, employing 80,000 people. This effort "produce[d] a crop worth approximately $1 billion US" (p. 357). Mariculture as a possible world food source should not, however, be overly stressed.

[21]Roger Revelle, "Present and Future State of Living Marine and Freshwater Resources," in Robert Repetto, ed., *The Global Possible, Resources, Development*

and the Next Century (New Haven and London: Yale University Press, 1985), pp.445-446.

[22]M. Grant Gross, *Oceanography, A View of the Earth*, 4th ed., op. cit., p. 395.

[23]Gross, op. cit., pp.297 and 258.

[24]Gross, op. cit., pp.297 and 258.

[25]Ibid.

[26]Satu Nurmi, "Issues and problems in the protection of the marine environment," op. cit., p. 207.

[27]See Dennis Pirages, *Global Technopolitics: The International Politics of Technology and Resources* (Pacific Grove, CA: Brooks/Cole Publishing Co., 1989), p. 45. He notes "[I]ndustrialization exacts a significant toll from the global environment because of toxic liquids, gases and solids that are by-products of industrial processes. Although in common language the word *disposed* is used in talking about these by-products, most of them are simply *dispersed* into the atmosphere and hydrosphere shared by all nations." Also, note *Protection and preservation of the marine environment, Report of the Secretary-General.* A/44/461, 18 September 1989, p. 25.

[28]GESAMP...: "Environmental capacity: An approach to marine pollution prevention." *UNEP Regional Seas Reports and Studies, No. 80.* UNEP, 1986.

[29]Ibid., p. iii.

[30]Ibid.

[31]Ibid.

[32]Ibid., p. i (emphasis added).

[33]Ibid., p. 2.

[34]Ibid., p. 15.

[35]Satu Nurmi, op. cit., p. 209.

[36]Ibid.

[37]C. H. Lawrence, *Medieval Monasticism, Forms of Religious Life in Western Europe in the Middle Ages*, 2nd ed. (London and New York: Longman, 1989), p. ix.

[38]Satu Nurmi, op. cit., p. 209.

[39]This paragraph is drawn from Satu Nurmi, op. cit., pp.209-211.

[40]Satu Nurmi, op. cit., p. 213. See also "Cleaning up the Seas, UNEP Environment Brief No. 5," (Nairobi: United Nations Environment Programme, n.d.).

[41]Satu Nurmi, op. cit., p. 212. See also *GESAMP: An oceanographic model for the dispersion of wastes disposed of in the deep sea* (No. 19), 1983, pp.1 and 4.

[42]I am reminded here, while beginning this track on the sources of marine pollution, of the preface to a classic work in medieval studies by Frederick B. Artz.

See his *The Mind of the Middle Ages, An Historical Survey A.D. 200-1500, Third Edition, Revised* (Chicago and London: The University of Chicago Press, 1980), p. vii. Artz perceptively noted on approaching his own task, "I approach the reader like St. Denis, with my head in my hands. I know that no specialist would write a book from Moses to Homer to Erasmus and Machiavelli. Yet, as Preserved Smith has said: 'The need is now, surely, for a return to synoptic writing. Here is a map of a large region, not a geological chart of a square mile or a plan of a single city' ... I know ... omissions will occur... But no two writers will agree on what to omit; any discussion of so vast and complex a subject must necessarily involve many lacunae."

[43]This paragraph is drawn from Gross, op. cit., p. 359.

[44]Ibid.

[45]The 10 are: Mediterranean Action Plan, Action Plan for the Protection and Development of the Marine Environment and the Coastal Areas of Bahrain, Iran, Iraq, Kuwait, Oman, Qatar, Saudi Arabia and the United Arab Emirates; Action Plan for the Protection of the Marine Environment and Development of Coastal Areas in the Red Sea and Gulf of Aden; Action Plan for the Caribbean Environment Programme; Action Plan for the Protection and Development of the Marine and Coastal Environment of the Eastern African Region; Action Plan for the Protection and Development of the Marine Environment and Coastal Areas of the West and Central African Region; Action Plan for the Protection of the Marine Environment and Coastal Areas of the South-East Pacific; Action Plan for Managing the Natural Resources and Environment of the South Pacific Region; Action Plan for the Protection and Development of the Marine and Coastal Areas of the East Asian Seas Region; and South Asian Seas Action Plan. See "UNEP Achievements, United Nations Environment Programme," (Nairobi: UNEP, n.d.), p. 5.

[46]*GESAMP: Health of the Oceans* (1982). See also "UNEP Environment Brief No. 5," (Nairobi: United Nations Environment Programme, n.d.)., p. 3.

[47]*GESAMP: The Health of the Oceans* (1982), p. 3 (emphasis added).

[48]"State of the Environment Reports, Emerging Environmental Issues." *Report of the Executive Director, United Nations Environment Programme.* UNEP/GC.15/7/Add. 3. 28 March 1989, pp.7-8.

[49]*The New York Times*, 16 August 1989, p. 4.

[50]"State of the Environment Reports, Emerging Environmental Issues," op. cit., p. 7.

[51]*The New York Times,* 16 August 1989, p. 4.

[52]"State of the Environment Reports, Emerging Environmental Issues," op. cit., p. 8. There have also been outbreaks in the North Sea in 1981 and in waters off the nations of southern Scandinavia in 1988.

[53]See *The New York Times*, 16 August 1989, p. 4, and *The Christian Science Monitor*, 3 August 1989, p. 3.

[54]See the comments of Prof. Roberto Marchetti of the University of Milan in *The Christian Science Monitor*, 3 August 1989, p. 3.

[55]The New York Times, 16 August 1989, p. 4. See also *GESAMP ....: Report of the Eighteenth Session. Paris, 11-15 April 1988. Rep. Stud. GESAMP, No. 33. 1988*, p. 36.

[56]This paragraph is drawn from a UNEP pamphlet, "Hazardous Chemicals, UNEP Environment Brief No. 4." (Nairobi: United Nations Environment Programme, n.d.), p. 1.

[57]Ibid.

[58]Ibid., p. 2.

[59]Ibid.

[60]Ibid. See also the box entitled "The diary of disaster," p. 3.

[61]See "Environmental events record" published in *UNEP News*, (Nairobi: United Nations Environment Programme, n.d.).

[62]Copy sent to author from Greenpeace, USA. Jim Wallette, "The International Trade in Wastes: A Greenpeace Inventory," (Washington, DC: 1988).

[63]*Africa News*, July 1988, in "The International Trade in Wastes: A Greenpeace Inventory," op. cit., p. 13.

[64]See John E. Carroll, "Introduction" in John E. Carroll, ed., *International Environmental Diplomacy, The management and resolution of transfrontier environmental problems* (Cambridge: Cambridge University Press, 1988), p. 1. "A transboundary environmental problem arises when all or most of the benefits of any pollution-emitting activity accrue to one nation, while all or most of the costs of that activity accrue to the nation across the border. It is the existence and the location of the border, therefore, and the imbalances or asymmetries in costs and benefits that result between two (or more) nations, which defines the existence of a transboundary environmental problem, and not necessarily the actual ... impact ... in question. And, indeed, since perception equals reality in the context of politics and diplomacy, if peoples of a nation perceive (believe) that damage is being done to them, a transboundary environmental problem exists, regardless of whether or not there is real environmental damage or cost. Perception thus becomes reality. Likewise, if transborder damage goes unrecognized by the populace affected, then there is generally no transboundary environmental problem."

[65]R. B. Clark, *Marine Pollution* (Oxford: Clarendon Press, 1986), p. 31.

[66]Satu Nurmi, op. cit., pp.214-215 notes: "Although land-based activities are responsible for the main flow of marine pollution, vessel-source oil is no doubt the most documented enemy of the marine environment. According to OECD [Organization of Economic Cooperation and Development] calculations, roughly 1.5 million tonnes of oil annually are discharged into the sea by shipping: 1 tonne for every 1,000 tonnes transported by sea. Estimates suggest that shipping is the source of less than half of all the oil reaching the oceans and that tanker accidents

account only for some 10% of the input from shipping. The largest source of oil discharges is of non-accidental origin and results from routine operations such as tank cleaning, deballasting and bilge discharges." See also M. O'Rathaille and P. Wiedemann, "The Social Cost of Marine Accidents and Marine Traffic Management Systems," *The Journal of Navigation* (January 1980) for a less sanguine view.

[67]Edward D. Goldberg, *The Health of the Oceans* (Paris: The UNESCO Press, 1976), p. 118.

[68]Ibid., pp.117 and 133.

[69]Ibid., p. 117.

[70]Ibid., p. 19.

[71]A. Golik, "Tar pollution in the Mediterranean Sea." *UNEP Regional Seas Reports and Studies, No. 91*, UNEP, 1988.

[72]Quoted in Ibid., p. 2. A. Zsolnay, B.F. Morris and J.N. Butler, "Relationship between aromatic hydrocarbons and pelagic tar in the Mediterranean Sea." *Environmental Conservation* 5 (4), pp.295-297.

[73]A. Golik, op. cit., p. 2 and see note 63 above.

[74]This paragraph is drawn from A. Golik, op. cit., p. 10.

[75]This paragraph is drawn from A. Golik, op. cit., pp.26-28.

[76]Studies have been conducted in Israel, the general western Mediterranean area, Cyprus and Turkey.

[77]J.L. Zhu and C.Y. Chan, Director of the Division of Nuclear Fuel Cycles and Waste Management, and a staff member of the Division, respectively, address these definitional questions in an article "Radioactive waste management: World overview." *IAEA Bulletin*, Vol. 31, No. 4, 1989, p. 6. The three terms are defined as follows: "*Low-level wastes* ... contain a negligible amount of long-lived radionuclides. Produced by peaceful nuclear activities in industry, medicine, research and by nuclear power operations, such wastes may include items such as packaged gloves, rags, glass, small tools, paper and filters which have been contaminated by radioactive material. *Intermediate-level wastes* ... contain lower levels of heat content than high-level wastes, but they still must be shielded during handling and transport. Such wastes may include resins from reactor operations or solidified chemical sludges, as well as pieces of equipment or metal fragments. Commercial engineering processes are being used to treat and immobilize these wastes; disposal in surface structures or shallow burial is practiced widely. Some countries have built or plan to build shallow repositories in rock formations on land or under the sea. *High-level wastes* ... arise from the reprocessing of spent fuel from nuclear power reactors through which uranium and plutonium can be recovered for re-use. These wastes contain transuranic elements, and fission products that are highly radioactive, heat-generating and long-lived. Liquid HLW has been effectively stored in tanks at specially constructed facilities. Before final

disposal and isolation from the biosphere, they require treatment and solidification. Spent fuel that is not reprocessed may be considered a high-level waste."

[78]D. P. Calmet, "Ocean disposal of radioactive waste: Status report." *IAEA Bulletin*, Vol. 31, No. 4, 1989.

[79]Satu Nurmi, op. cit., pp.217-218.

[80]For a limited and preliminary report see Alf Larsson, "State-of-the-art report on radio-active waste disposal." *IAEA Bulletin*, Vol. 31, No. 4, 1989. This article discussed the forthcoming IAEA report on "State-of-the-art" radioactive waste disposal issues such as hydrology, geochemical concerns, engineering requirements, design concepts and performances of disposal systems.

[81]This paragraph is drawn from *The Times-Picayune* (New Orleans, LA), 9 March 1990, p. 7 from *The Baltimore Sun*.

[82]GESAMP ...: The state of the marine environment. *UNEP Regional Seas Reports and Studies, No. 115*. UNEP, 1990.

[83]Ibid., p. 39.

[84]Ibid.

[85]Ibid.

[86]France, of course, still continues as of this writing (Winter 1990) to conduct atmospheric tests in the south Pacific region.

[87]GESAMP ...: *Regional Seas Reports and Studies, No. 115*, op. cit., p. 39. See also GESAMP ...: Thermal discharges in the marine environment. *UNEP Regional Seas Reports and Studies, No. 45*. UNEP, 1984, pp.20-21.

[88]GESAMP ...: *UNEP Regional Seas Reports and Studies, No. 45*, op. cit., p. 20.

[89]D. P. Calmet, op. cit., p. 47.

[90]Each PBq=10^{15} becquerels. Becquerel rays, according to *The American Heritage Dictionary of Science*, compiled by Robert K. Barnhart and published by Houghton Mifflin in 1986, are (p. 63) "the invisible rays given off by radioactive substances (replaced by the more specific terms, *alpha rays, beta rays* and *gamma rays*). Named after Antoine H. Becquerel, who first reported these rays."

[91]GESAMP ...: The state of the marine environment. *UNEP Regional Seas Reports and Studies, No. 115*, op. cit., p. 42.

[92]D. P. Calmet, op. cit., p. 47.

[93]GESAMP ...: The state of the marine environment. *UNEP Regional Seas Reports and Studies, No. 115*, op. cit.

[94]GESAMP ...: Interchange of Pollutants between the Atmosphere and the Oceans, 1980. *Rep. Stud.* (13), p. 1.

[95]Ibid., (emphasis added), p. 2.

[96]Ibid.

[97]See the brochure "A Study of Global Change" published by the *International Council of Scientific Unions*, Paris (n.d.).

[98]GESAMP ...: The state of the marine environment, op. cit., pp.34-35.

[99]This paragraph is drawn from *Toward an Understanding of Global Change* (Washington, DC: National Academy Press, 1988), p. 1.

[100]See Robert E. Dickinson, "Climate and Hydrological Systems" in *Toward an Understanding of Global Change*, op. cit., p. 130.

[101]"A Study of Global Change" published by the *International Council of Scientific Unions*, Paris (n.d. or page).

[102]As of this writing (Winter 1990), the Bush Administration has yet to accept the implications of increased carbon dioxide emissions into the atmosphere and likely consequences of this for ocean levels. See *Science*, 24 November 1989, pp.992-993. The first two paragraphs tell the story: "A slim, unrefereed report that many scientists have dismissed as biased and misleading is said to be at least partly behind the White House's recent temporizing on climate change... And that is causing consternation among climatologists and other greenhouse experts, who are dismayed that this 35-page document by the George C. Marshall Institute, a Washington, DC think tank–rather than one of the massive, carefully researched and reviewed expert reports of the past several years–seems to be holding sway in the upper echelons of the Administration."

[103]G. Sestini, L. Jeftic and J.D. Milliman, "Implications of expected climate changes in the Mediterranean region: an overview." *UNEP Regional Seas Reports and Studies No. 103*, UNEP, 1989, p. i and p. 1 of Introduction.

[104]This paragraph is drawn from Ibid., p. 7.

[105]Ibid.

[106]Ibid. Other statistics from the UNEP report are equally alarming. "Some 85% of the waste waters from the larger cities is drained untreated into the sea. This ... does not take into account river-borne transport, which accounts for some 40% of all pollution. *A significant share of modern equipment for the purification of waste water does not [even] function properly*" (emphasis added).

[107]See T. M. L. Wigley, "Future climate of the Mediterranean Basin, with particular emphasis on changes in precipitation," in G. Sestini, L. Jeftic and J.D. Milliman, eds., (in preparation, n.d.) quoted in G. Sestini, et al., "Implications of expected climate changes in the Mediterranean region: an overview." *UNEP Regional Seas Reports and Studies, No. 103*, op. cit., p. 14.

[108]G. Sestini, et al., *UNEP Regional Seas Reports and Studies No. 103*, op. cit., p. 14. "Regional changes in all climatic variables will occur. So far General Circulation Models ... cannot simulate the regional details of today's climate reliably. Projections using these models therefore must be treated cautiously" [i.e., the general circulation models]. This problem has been a special concern with those working in the theoretical aspects of Chaos science. See Ian Percival, "Chaos: a science for the real world." *New Scientist*, 21 October 1989, p. 14. For further information, consult Eric B. Kraus, "The Role of the Ocean in Climate

Fluctuations," in S. Fred Singer, *The Ocean in Human Affairs* (New York: Paragon House, 1990).

[109]G. Sestini, et al., op. cit., p. 20.

[110]Ibid., p. 21

[111]Ibid.

[112]Ibid., p. 22.

4

The Contributions of Intergovernmental Organizations

Introduction

This chapter addresses the topic of selected intergovernmental organizations (IGOs), their role in the ocean degradation question and the function of the United Nations as a useful instrument in the solution of pollution problems. The goal is to examine the work of four organizations involved in the question of ocean pollution on a planetary scale. The four are: the United Nations Environment Programme (UNEP), the United Nations Educational, Scientific and Cultural Organization (UNESCO), the International Oceanographic Commission (IOC) and the World Meteorological Organization (WMO). These were chosen for the scope and intensity of their involvement in the ocean pollution issue. All, by the very nature of their function and activity are intimately involved on a day-to-day basis with the issues examined in this chapter. This participation expresses itself through working panels, special conferences, study groups and numerous other activities which involve the IGOs, usually in major ways, as the primary focus points for the dissemination of knowledge and the building of consensus within the relevant scientific communities for problems at hand. The transformation of the knowledge engendered through these inter-disciplinary activities into effective governmental action by the world's nation-states is, however, another question entirely; and while some of those problems will be discussed here, the main conclusions will have to wait until the final chapter of this book. Research for this section was drawn primarily from personal interviews conducted by the author in the Spring of 1989 (with officials at the headquarters of the respective agencies, the only exception being that of the United Nations Environment Programme in Nairobi), and examination of various reports and documents of relevant United Nations agencies.

The United Nations itself, an organization of more than 169 member states with a secretariat and other major organs (the General Assembly, the Security Council, the Economic and Social Council, the Trusteeship Council and the International Court of Justice at The Hague), as opposed to the semi-independent international agencies which nominally belong to what has been called the United Nations system,[1] is a unique institution. Nothing has existed quite like it, the League of

Nations notwithstanding.[2] The United Nations (UN) embodies the ideals of universal membership, international law and most importantly functions as *the* world forum where not only all nations may speak (however offensive this may be to some, however obtuse the arguments may be framed to others–and, of course, many times they are), but where *global issues* may be raised and discussed. The latter include such problems as war and peace, environmental degradation and economic interdependence and development. The UN has been a thorn in the side of many, an integral part of international politics in its early years, a dis-used and abused organization in its most recent past and in the last few years a renascent IGO—but one still fully under the tutelage of its member-states.[3] Of course, the member-states most important in this question of leadership and effectiveness are the five permanent members of the Security Council.

The UN has been important for defusing many potential threats to world peace, but in the last several years it has become more important because of the urgency surrounding global environmental questions. Pollution knows no boundaries—national borders are irrelevant to acid rain, or ozone depletion of the planet's atmosphere, or chemical poisoning leaking into lakes and rivers and estuaries which then drain into the ocean. One of the most important functions of the organization is to be an educator. It has little scope in which to act independently from the member-states who fund its budget and control its activities.[4] Nevertheless, as an actor in the international system and as the largest and most influential intergovernmental organization, the UN has a unique hold on the attention of national governments. Nearly every state is a member; those few which are not usually have official observer status (Switzerland, the Holy See, and the two Koreas are the most prominent examples). The UN is also the central clearing-house or sponsor, or at times organizer, for the myriad of multinational environmental study and recommendation efforts in which nearly every actor of the international system is now engaged.

This is not to say the United Nations organization is the main locale for environmental studies. That would ignore the efforts of national governments and their respective academies of science,[5] as well as other groups of experts engaged in these studies–including, of course, the efforts of scholars working on many of the same questions in nations spread out all over the planet. These national institutes, some several hundreds of years old, are vigorously involved in such research. So, too, are the literally thousands of universities throughout the world which are feverishly studying environmental questions. But what differentiates all these efforts from those held under various United Nations fora is the unique nature of many of the programs administered or jointly sponsored by the organization. In addition, the organization possesses universality in its membership which is a factor that should not be overlooked. This organization then, the United Nations, is made up, through its member-states, of all the peoples of Planet Earth. This phenomenon is not replicable on a global scale by individual national governments.

UNEP

The United Nations Environment Programme (UNEP) was established as a result of initiatives undertaken at the 1972 UN sponsored Conference on the Human Environment held in Stockholm. This first ever held truly global meeting to discuss and recommend courses of action for the total environment of Planet Earth (air, land and water), and made up of national governments and intergovernmental organizations, along with international nongovernmental organizations (INGOs), was convened at a time of increasing concern world-wide over anthropogenic changes altering the planet's life-support systems. UNEP was established by a General Assembly resolution, as a specialized agency of the United Nations, "to serve as a focal point for environmental action and coordination within the UN System."[6] It is not, therefore, an "operating agency." It has no large staff, works within a modest budget and was envisioned principally to serve as a "facilitator." The goal was to create a secretariat that would serve the international community as "a small focus of global know-how, charged with stimulating effective global environmental action through the coordination of independent activities, leadership and efforts to catalyse and stimulate action by others... [It] was not funded to address major environmental concerns by itself..."[7]

A. Leroy Bennett has analyzed the organization structure of this unique UN agency. He has noted, that

> The permanent machinery created by the General Assembly includes a small Environmental Secretariat ...; a fifty-eight member Governing Council to coordinate environmental activities and to promote governmental cooperation; an Environment Fund raised by voluntary contributions with a target of 100 million dollars for the first five years; and an Environment Coordination Board as a subsidiary of the intersecretariat Administrative Committee on Coordination to foster cooperation among the secretariats of the United Nations and its specialized agencies in carrying out environmental projects.[8]

The circumspect nature of UNEPs authorization from the General Assembly, however, should not be confused with the level of activity pursued by the agency. With limited monies ("UNEP operates on a two year budget averaging approximately $29.5 million [US] annually. This figure represents both convertible and non-convertible currencies."[9]) the organization sponsors, funds or participates in a host of environmental activities. These include international conferences such as the 1987 Montreal meeting which resulted in adoption of a Protocol on Substances that Deplete the Ozone Layer, as well as the more day-to-day routine scientific panels and working groups.[10]

Again, it is not possible within the confines of this chapter to list or discuss all the relevant reports or activities. The numbers are far too voluminous for that. However, the wide scope of UNEPs activities in one area, the Regional Seas Reports and Studies, can perhaps be indicative of the efforts involved. More than 115 of these reports have been released since 1982. They are of high quality and include recognized experts in their respective fields. The first substantive research report was undertaken as a cooperative effort with the United Nations Industrial Development Organization (UNIDO). UNEP and UNIDO released a "Survey of marine pollutants from industrial sources in the West and Central African region." Other reports have studied levels and effects of marine pollution in the south-east Pacific region, and the state of marine pollution in the Caribbean region. More recent panels have examined "Coastal and marine environmental problems of the United Republic of Tanzania"—a wide-ranging investigation into the problem areas and capabilities of the Tanzanian marine science program.

One of the important and potentially overtly contentious political acts undertaken by UNEP concerns the establishment and operation of the Global Environment Monitoring System (GEMS).[11] This program utilized monitoring stations throughout the world and is an ambitious attempt to predict environmental threats by construction of a sophisticated database system. "The programme collates a huge variety of environmental statistics, including information from satellites and monitoring agencies on the ground. It tracks pollutants as they travel through the air and sea, and keeps an eye on shifting patterns of land cover, and the state of the atmosphere, the climate and endangered species."[12] The idea behind the system, of course, is either to predict upcoming events, or at least to significantly raise the levels of knowledge concerning the nature of the threats posed.

The director of GEMS has noted the potential for serious political problems for the program if their forecasts turn out to be wrong. "Supposing a government had invested a huge amount of money in an attempt to ward off some environmental disaster ... and we had got the forecast wrong."[13] This would certainly be a serious matter. Of equal concern, however, if one is ruminating on possible future scenarios, is the likelihood of GEMS making a prediction or issuing a strong warning and a government ignoring it because of short-term economic or political costs. This could be done quite easily by national authorities disputing the data itself or the interpretation given to it by GEMS. Contrary to the general or common wisdom, science is not a "cut and dry process," i.e., it is not a simple matter of looking at the data (usually a very large amount in these matters, and often confusing as well) and pronouncing in an olympian manner that the question falls to either one side or another. From the nascent idea, to the design of a project, thru the conduct of preliminary experiments and later ones, constantly interpreting data and "concluding" with a set of final results—all this is inherently a complicated process. Politicians (and others), therefore, have a ready-made occasion in which to question the methods employed or the conclusions reached and to avoid taking definitive (or even significant) action.[14]

UNESCO

The United Nations Educational, Scientific and Cultural Organization (UNESCO) is the major component for marine research in the UN system. It sponsors or directly undertakes many programs of study addressing a wide variety of issues affecting the global ocean.

The Division of Marine Sciences is the primary organ of marine science research within UNESCO. The main focus of the division is the concentration on the development of scientific knowledge, specifically concerning coastal regions and the management thereof (it will be remembered that UNEP has primary responsibility for regional seas, although UNEP is not a management but a coordination agency).[15] The division is of critical importance in the efforts of many related entities (whether they are within the parameters of IGOs such as UNESCO, or INGOs that cooperate closely with UNESCO, such as the International Council of Scientific Unions—ICSU) attempting to understand the basic scientific questions associated with biogeochemical processes in the ocean.

The 1988-1989 draft program of the division for training and research in regard to coastal management stated the following: "The aim of this programme is to promote the development of the marine sciences as a basis for the management of the coastal environment, the rational exploitation of marine resources, and the advancement of knowledge."[16] Some of these efforts will be discussed in this section. Unfortunately, the division is involved in so many activities that an analysis of them all would prove unwieldy.[17]

The Intergovernmental Oceanographic Commission (see below) is also part of the UNESCO structure charged with addressing certain ocean related issues in the wider system of UN agencies, organizations and other affiliated bodies.

> Its purpose is 'to promote scientific investigation with a view to learning more about the nature and resources of the oceans through the concerted action of its members.' It further functions as *the* co-ordinating body within the United Nations system for marine science and related activities. Co-operative scientific investigations of the ocean as well as world-wide ocean

> services, combined with a programme of training, education and mutual assistance, form the chief components of IOC's work.[18]

The Division of Marine Sciences (DMS) is primarily concerned with the "UNESCO Major Interregional Project on Research and Training leading to the Integrated Management of Coastal Systems" (COMAR). Coastal areas are critical in the study of the degradation of the marine environment. More than two thirds of the population of the planet live in or near coasts, estuaries or deltas (within 50 miles).[19] Ocean pollution is most serious in coastal areas. DMS and the COMAR project work closely with the International Council of Scientific Unions (ICSU) regarding various regional programs in Latin America, the Caribbean, Asia, the Pacific, the Mediterranean, the Red Sea and Europe.[20]

COMAR is also coordinated with the work and activities of the IOC. The medium term aims of the project are varied, and include efforts "... to promote research and training on key aspects of coastal marine systems. The ultimate goal is to provide a reliable scientific basis for the management of the coastal marine environment and its resources."[21] The activities undertaken in this program along the planet's coastlines are intended to form the basis for a system of management that can protect, regulate and utilize the resources which marine coastal systems contain.[22] A COMAR pamphlet, published by UNESCO, notes

> Generally, marine coastal systems are highly productive. They provide nursery grounds and shelter for a number of marine species. Their biological communities play a major role in the world's fisheries and may also constitute gene pools. More than four fifths of the fishery catches come from coastal regions... Living and non-living resources of these environments yield a wide variety of products ranging from fuel to construction material and from food to medicines... [M]ineral resources are [also] ... exploited in the coastal zone.[23]

COMAR has a "core project" and associated regional projects. The coastal systems which are of concern to the Interregional Project are under tremendous assault due to anthropogenic intervention. Traditional means of human interaction with this environment, including historical exploitative practices, are changing due to a variety of pressures: increased population, technological improvements relating to the efficiency of resource extraction and the concomitant by-product of increased polluting substances. These issues are being studied to develop basic knowledge of the problem and possible solutions.

The rationale behind COMAR holds that these transformations already occurring within the coastal environment need to be guided and based on the scientific principles that govern and regulate the natural systems present. "Decisions regarding alternative uses of coastal systems, and of their resources including their conservation for agri- and aquacultural ends, should be based on a proper understanding of the functioning and mutual interaction of the natural systems concerned. If not, long-term benefits may be sacrificed for short-term gains."[24] The last sentence, it should be noted, is an understatement.

The core project "... is aimed at reviewing and promoting research and training on the main aspects of the various coastal systems." These include:

1. the ecological characteristics and functioning;
2. the interactions and exchange of energy and material; and
3. the relations with sea and land.[25]

DMS carries out this work through a sophisticated and complex interlocking network of IGOs and INGOs. Regular meetings of scientific experts provide the basis for COMAR's activities. This system relies heavily on research groups, special workshops and panels which exist ostensibly to improve the management decisions which are at the center of the DMS mission.

Regional COMAR activities are concerned with "... reinforcing the capacities of the countries [concerned] in building up a base of scientific knowledge for the rational management of the coastal zone." The regional efforts also maintain informational and organizational links with the Core project. The primary concerns of the regional activities also include matters relating to:

1. the diversity and complexity of the interaction between human societies and the coastal marine environment;
2. the limitation of natural resources and destabilization of natural coastal marine ecosystems;
3. the need for new knowledge and adequate dissemination of existing knowledge; and
4. the limitation of means and manpower [as these effect the plan's implementation and success].[26]

As in the case with all UN programs and activities, the agencies concerned cannot "force" policies or guidelines upon nation-states. They can only work cooperatively and recommend suggested action. If a relationship not only of trust and respect has been created, but also if the full and active involvement of the states in question have been sought and cultivated, then the chances for success are greatly enhanced (but not ensured). In the limited role which DMS and COMAR have devised, success has been noticeable.[27] This has greatly depended upon the practicality with which the program has been conducted. As an example, one can note the description of the program's operation provided by COMAR. "The modality of project implementation in each region depends on the types of activities selected for the regions. The type ... depends, in particular, on the existing *specialized manpower and facilities*, as well as on the *level of participation* and *financial commitments* of the countries and the international organizations."[28]

IOC

The Intergovernmental Oceanographic Commission (IOC) is the most active and influential component of the UN system, after the United Nations Environment Programme, engaged in oceanic research; including studies of ocean pollution. As the leading body within the international system for cooperative intergovernmental oceanography studies, the IOC sponsors or directs numerous programs that examine basic processes of ocean climate, circulation and pollution. The 1984-1985 report of the Director-General of UNESCO noted that

> The major global scientific programmes of the intergovernmental Oceanographic Commission ... include studies on the relationship between ocean processes and climate; ocean science in relation to living and non-living resources; and the study of the state of the health of the ocean, including the monitoring of marine pollution. These programmes are undertaken in close co-operation with the scientific communities concerned and organizations of the United Nations system.[29]

Three main program areas exist within the IOC: ocean science, ocean services and training, and education and mutual assistance.[30] The first and most important of these programs is the global investigation of marine pollution.

The Global Investigation of Pollution in the Marine Environment (GIPME) is the lead program in the ocean science area. In the history of the IOC[31] "[t]he protection of the marine environment ... was early recognized to be an important issue."[32] As early as 1960, UNESCO convened an intergovernmental conference on oceanic research to request "... states to take, without delay, all steps in their power to prevent pollution of oceans and seas by radioactive and other harmful agents."[33] These efforts led to the establishment of an "... IOC *Working Group on Marine Pollution* with the task of studying how the Commission could further national and international endeavours in this ... field." Further details on the basic story can be found in the relevant history, and are not important to the points raised here.[34]

By 1972 GIPME was operating. "It was composed jointly of scientists representing Member States or relevant organizations of the UN system who were actively involved in the conduct of marine pollution research programmes." In

1976, at the instigation of GIPME, a report was written and released on the health of the global ocean which analyzed ocean pollution based on five modes: halogenated hydrocarbons, radioactivity, heavy metals, petroleum hydrocarbons and litter.[35] This report provided guidance to the IOC in its continuing efforts to monitor ocean pollution.

Ocean services are the second major activity of the commission.[36] The commission is actively involved in the exchange of relevant information to interested scientists, governments and the public at large. The information concerns the wide variety of knowledge which flows constantly from oceanographic research and is in a constant state of up-dating and adaptation. The commission performs, in this role, the function of a sort of communications clearinghouse for professionals and professional studies. As Roll has noted:

> It is the characteristic of such a "service" that specified information is provided to interested users in a more or less regular fashion or upon special request or whenever such information is needed to avoid damage and destruction...

The activities of the commission in this regard are discriminatory in the sense that objectives of the ocean services component are divided into two parts: first, a responsibility to the scientific community; and second, to governments and interested publics. This discrimination reflects the technical nature of the commission's work, and illustrates the complexity of issues which scientists in this field first must understand before the benefit of their expertise and study can be shared.

The ocean services component embraces several programs, of which a discussion of them all would be quite lengthy. However, one which will be addressed is the Integrated Global Ocean Station System (IGOSS). IOC works closely with the World Meteorological Organization (WMO) in the administration and operation of this network. The overlap here of the work of the two organizations is not atypical of ocean science and pollution studies within the UN system. The following section on the work of the WMO and ocean pollution will go into more detail. For now, however, the analysis in this section will be confined to the IGOSS network.

The IGOSS system is a "network of fixed oceanographic ... [observation] stations ..."[37] in a world-wide meteorological network. "The purpose of IGOSS is to promote, develop and co-ordinate the international machinery necessary for the timely global acquisition and exchange of ocean data; and ... [its] dissemination ..." [to interested parties].[38] Utilizing time-series data and summary observations, IGOSS provides a basis "... for monitoring and modelling ocean processes with the aim of understanding the nature of the ocean."[39] The advantages of this system for the study of ocean pollution are clear. Timely information provided by the network can help protect the natural resources in the ocean, can identify effluents and garner information for the scientific management of waste, and can also monitor the distribution, transportation and decomposition of pollutants.[40]

IGOSS compliments other work within the IOC.[41] In 1970 a Group of Experts on Long-term Scientific Policy and Planning (GELTSPAP) was established for the purpose of advising IOC on its own Long-term and Expanded Programme of Oceanic exploration and Research (LEPOR). GELTSPAP pointed out the suitability and necessity of "... identifying the elements of a desirable and feasible marine pollution monitoring system." Monitoring was defined as "a system of continued observation, measurement and evaluation for defined purposes." These defined purposes were to include:

1. providing early warning of significant environmental changes to induce consideration of protective measures; and
2. enhancing quantitative knowledge of the marine environment, including the means by which dynamic balance is maintained in marine ecosystems, as a basis for managing marine resources.

Responsibility for the implementation of this mandate was given to the GIPME and IGOSS programs. In addition, in order to realize this mandate IOC developed a Marine Pollution Monitoring Programme and a pilot project. The project utilized available technology to monitor pollution indices. A plan was drawn up for this and a Marine Pollution Monitoring Pilot Project (MAPMOPP) was established. It focused only on the petroleum aspects of marine pollution, however. "The main objective of the ... [p]roject was to determine the feasibility of developing and operating such a programme on a global scale, with the ultimate goal being to obtain information on the distribution and dynamics of petroleum in the world ocean."[42] Roll found a wide brief for the operation of MAPMOPP, albeit of limited scope. It included

1. visual observations of oil slicks, petroleum residues and other floating pollutants (e.g. plastic materials);
2. collecting surface samples for later inspection in laboratories with respect to particulate petroleum residues (tar balls);
3. tar sampling on beaches; and
4. taking water samples for later analysis in laboratories with regard to petroleum hydrocarbons dissolved and dispersed in sea water.

The IOC is also active in cooperative efforts to improve the oceanographic capabilities of developing states. This category comprises a program known as

TEMA–Training, Education and Mutual Assistance in the marine sciences.[43] Such cooperative efforts have long been an integral part of the commission's work. "[T]his is understandable with a body like IOC, established to promote international co-operation and composed of Member States in very different stages of development in the field of marine science." Roll further noted, "[i]mproving the marine scientific capabilities of the less developed states is an absolute necessity if sufficient participation in IOCs programme is to be achieved."

The first "comprehensive discussions" on TEMA occurred in 1964. Subsequent efforts during that decade and the early years of the 1970s resulted in the adoption of a governing strategy for addressing the needs of developing states in the field of oceanography studies and the efforts the IOC could make in order to help upgrade national capabilities. Ten main goals were adopted. The three most important were: 1) "training of marine science administrations;" 2) "translations of important marine science texts;" and 3) the establishment of "regional training centres."[44]

Progress has been limited, however. The needs have grown more compelling. A 1985 IOC report noted that "... marine science and ocean services capabilities of developing states can be enhanced through active participation ... in well-defined national, regional and global programmes. A major programme is needed simultaneously to assist in building up facilities and trained and experienced personnel, and in strengthening national organizational infrastructures."[45] Because of changes in international law, and particularly because of the three United Nations conferences on the Law of the Sea,[46] the position of states (especially developing states) with regard to maritime jurisdiction have become more complex and interrelated not only with neighboring states but also to the wider international community as well. One small, but nevertheless significant component to this problem, was illustrated in the summary statement of the 1985 report. It noted "... [e]xtended maritime jurisdiction by coastal states has created a situation where self-reliance in marine science is needed to ensure that their new rights, especially to control resource exploration and exploitation, and the conduct of research, can be exercised and that they can benefit from this new situation for socio-economic development."[47] As matters stand at the present time, coastal developing states do not have the necessary resources (human and technical) to manage the enlarged areas of maritime national jurisdiction. They are not receiving sufficient help from the international community to solve this problem.

WMO

The World Meteorological Organization (WMO) is another specialized agency of the United Nations system and is, in addition, because of its technical capabilities, a pivotal servitor to the international community. It sponsors a host of vital scientific endeavors which are increasingly important to nation-states as heretofore "constant" environmental factors are undergoing radical transformation.

The WMO is a complex organization, containing many different kinds of research programs relating to the overall problem of studying the atmosphere, atmospheric conditions and particularly, of course, weather patterns. This section will focus, however, on only one such program—the World Ocean Circulation Experiment (WOCE). This is a large-scale, multi-disciplinary effort utilizing the resources of many nation-states, and various agencies and affiliated bodies of the United Nations system along with international nongovernmental organizations. Research on ocean circulation is important for ocean pollution studies because, quite obviously, pollution in water is transported in various ways around the globe. No part of the global ocean is immune to changes in other sectors.

The activities of WMO are varied, and it is useful to first review some of its formal responsibilities.[48] The WMO has seven major scientific and research programs (under the aegis of WOCE). The relevant program for attention here is the World Climate Program (WCP). It "... promotes the use of climate information to assist economic and social planning and development, the improvement of the understanding of climate processes through internationally coordinated research, and the monitoring of climate variations or changes in order to be able to warn governments of climate impacts that may significantly affect human welfare and activities."[49] The WCP contains four major research components: 1) the World Climate Data Program; 2) the World Climate Applications Program; 3) the World Climate Impact Studies Program, and 4) the World Climate Research Program."[50] And the "WOCE is the major oceanographic component of the WCRP [World Climate Research Program] Third Stream, dealing with decadal and longer period variability."[51] This Third Stream is the vital element in WOCE for understanding the long-term impacts of anthropogenic intervention in the marine environment.

The rationale behind WOCE is both ambitious scientifically and challenging to the organizations sponsoring the work: ICSU, WMO and the IOC. G. A. McBean and A. D. McEwan have noted the importance of global climate studies.

> Global climate change and its effects are emerging as the most pressing environmental problem for ... [humankind]. Although advances in computer power have put within our vision the possibility of reliable climate forecasts, it is now realised that successful simulation of the broad evolution of global climate beyond a few years will require the inclusion of the dynamics of the world oceans through their full depth.[52]

The WOCE is, therefore, an attempt to understand at the basic level the processes involved in planetary ocean circulation for decades ahead, not just two or three or five years into the future. The program will need the highest yet achieved quality of global oceanographic data. Experiments will be conducted whose goal is the acquisition of "unprecedented detail" in the data "which defines the circulation of the oceans and the exchanges that occur within and between them."[53] Finally as these authors conclude in their concise Forward, "Apart from being the largest coherent oceanographic experiment ever contemplated, WOCE is an essential step toward the rational description of climate and its evolution. Ultimately, benefits will accrue to nations, both individually and collectively. However, the success of WOCE is dependent upon international oceanographic participation and cooperation to an unprecedented degree."[54] Given the development of humankind's industrial apparatuses and the global impact these processes have introduced into the biosphere, the WOCE is a logical step to compliment and understand the processes involved in the circulation patterns of the Global Ocean.

WOCE is composed of three core projects. Core Project 1 is designated "The Global Description." Core Project 2 has been titled "The Southern Ocean," and Core Project 3 is named "The Gyre Dynamics Experiment."

Project 1 is the first step in the efforts of the WOCE to realize its mandate of determining how extensively climate can be predicted, and the impact humanity has on the planet's climate system. The goal of the Global Description project is to "... develop models useful for predicting climate change and to collect the data necessary to test them."[55] As such, the project has very broad or general objectives. They are also at the same time, however, quite specific and rigorous. Empirically, the idea is to achieve an overall understanding of circulation in the global ocean and the variability that is present within the general parameters. These factors will then be compared to the necessity of atmospheric forcing,[56] which will require changes in the way the scientific community understands the whole concept of ocean dynamics.

Observation and modelling are at the heart of this process. Core Project 1 contains 10 major observational components, of which the three widest in scope are the following:

1. large scale, full-depth, global hydrographic/tracer survey covering the entire ocean once;
2. a global deep float release employing a mixture of pop-ups and acoustically-tracked floats; and
3. satellite altimetry calibrated with a sparse, global sea-level network.[57]

This is global scientific research (Core Project 1). It is "big science,"[58] with the laboratory the planet itself. Core Project 1 will lay the groundwork not for the control of the planet's climate system (although knowledge and information will be generated which will assist at some future point perhaps in this end), but in the efficacious management of weather—a millennial dream of humanity.

Core Project 2 investigates what has come to be called by climatologists "The Southern Ocean," or what might more widely be known amongst non-specialists as the ocean area surrounding the continent Antarctica itself. Whatever the opaqueness of the term among the informed public, the Southern Ocean is the water basin surrounding the continent. Yet, as D. W. H. Walton has noted, "Antarctica is both a continent and an ocean. The continent, comprising almost 10% of the land surface of the globe, is inextricably linked to the vast extent of the Southern Ocean surrounding it."[59]

This "southern ocean" is a critical component of the World Ocean. Approximately "... 55 to 60%" of the world ocean "... owes its characteristics to Southern Ocean processes."[60] It is the depository for a major portion of the heat residing in the world ocean (as a heat sink) and is the primary connection between the three major "ocean basins," i.e. the Atlantic, Indian and Pacific. How all these processes fully interact is, of course, unknown. Questions such as water mass formation and spreading, as one example, are some of the concerns addressed by project 2.

Project 2 is an attempt to understand the complex role played by the southern ocean in the planetary climate system. This project, as well as the overall WOCE, seeks to fundamentally augment and enhance the quantitative understanding of this sub-system and its relationship to the larger patterns of climate and weather. "The effort [is planned to] ... contribute to the construction, improvement and verification of large-scale ocean models that will include ... modelling of the sea ice cover" (naturally, an essentially element in evaluating Antarctic climate). The mode of data acquisition will utilize remote sensing satellites, arrays of moored instruments, and various drifting and floating devices for a comprehensive survey of the region.[61]

Finally, Core Project 3 (The Gyre-Dynamics Experiment) is designed as an evolving series of tests, the exact nature and number of which are purposefully not yet known, that "... will study the processes that must be better understood if decadal climate predictions using models are to be made by the end of [the] WOCE [approximately 1995]."[62] In other words, the plan acknowledges that changes will have to be incorporated into any programs that are drawn up because the state of knowledge in this particular field is so far limited and tentative. The Scientific Plan itself notes it should be possible to study identified processes in one ocean basin (for instance, eddy-resolving models) that should then allow researchers to take that knowledge and extend it to models for the global ocean. That is the plan. "A major feature of these advances would be to explain and simulate the dominant modes of variability over a broad spectrum of spatial and temporal scales and to develop parametrisations of important processes that would have general applicability in ocean modelling." The WOCE report, however, also notes the inherent difficulties in this stage of the research.

> It is difficult to identify the particular processes which must be better incorporated in ocean models in order that these models can better represent the existing state of the ocean, let alone be used for decadal climate prediction. As more is learned about the dynamics of the ocean and models are developed with ever increasing resolution ... new problems will arise and new aspects of ocean physics [will] need clarification. Thus, the experiments put forward for Core Project 3 ... will need to be augmented and modified during the WOCE Intensive Observation Period.

Preliminary operational plans exist for beginning the study of the proposed "... augmented basin coverage" by focusing on the Atlantic for a period of five years. This area was chosen for its relative ease in operationalizing the goals of the program. "The principal reasons [for choosing the Atlantic] ... are its strong manifestations of diabatic processes (water mass conversion), size, previous measurements and its proximity to major shipping bases." The size factor is advantageous because fewer measurements are necessary for requisite resolution than, say, the Pacific. Historically, many studies have been done on Atlantic meteorological processes which can be used in the development of the program and its experiments. And the advantage of access to ports is obvious.

Conclusions

Nation-states are not the only actors within the international system that can conduct high-level, sophisticated, well-organized and complex scientific research. Nation-states also are not the only organized entities that can sponsor or host, or actually carry out, the work of investigating global environmental problems. Finally, the utilization of the nation-state framework as the essential element in addressing the ocean degradation issue is not primarily, nor necessarily, the most *a priori* logical or efficacious means in which to confront the issue of global pollution.

None of the above observations are novel, nor are they particularly shattering in their identification. Informed observers would, of course, understand that it is perfectly feasible, and clearly within the realm of their capabilities (given minimum standards of funding and organizational support), for IGOs to create structures that facilitate global cooperative efforts. This is true not only in undertaking basic research, but also in administering limited and very defined programs in ocean science. The UN system has been engaged in this type of activity or "business" for over forty years. However, the obvious often has great heuristic value; not to mention the ability to reveal what might be for some unpleasant evidence: that the nation-state is not some sort of quasi-religious totem that, by touching it and offering obsequious reassurance, spreads it efficacious or inevitable "mana" in the form of satisfactory solutions to formidable social problems. This theme will be further explored in the final conclusion to this book.

This chapter examined the contributions of four selected IGOs; including, parenthetically, their attempt to contribute to education and consensus building on a world-wide scale. The second intention of the chapter was to demonstrate to the reader the ecumenical nature of problems affecting the Global Ocean, the interrelatedness of the issues involved, and the activities of the international community—expressed through their delegated agencies—to coordinate research and action on the national level.

Chapter 5 will proceed by examining the work and activities of two selected international nongovernmental organizations (INGOs) and their efforts in the field of ocean pollution and species protection.

Chapter 4 Notes

[1]See Amos Yoder, *The Evolution of the United Nations System* (New York: Taylor and Francis, 1989).

[2]For brief histories of the League, see Amos Yoder, op. cit., pp. 8-25; Peter R. Baehr and Leon Gordenker, *The United Nations, Reality and Ideal* (New York: Praeger Publishers, 1984), pp. 1-16; and Werner J. Feld and Robert S. Jordan, with Leon Hurwitz, *International Organizations, A Comparative Approach*, 2nd ed. (New York: Praeger Publishers, 1988), pp. 42-45.

[3]Here Amos Yoder's introduction is quite useful and succinctly stated. See his *The Evolution of the United Nations System*, op. cit., pp. 1-7.

[4]Some organs of the United Nations do have limited scope to temporarily "flout" the will of even the more powerful members. The Secretariat, particularly the Secretary-General, has at times in the past pursued policies that were viewed, at least by some permanent members of the Security Council, as an independent or dangerously separate course of action. This resulted in either outright hostility from one permanent member—the Soviet Union in response to Trygve Lie's support of the position of the United States in the Security Council with regard to the North Korean invasion of South Korea—or the strained relations Dag Hammarskjold endured with the United States over loyalty questions vis-a-vis suspected communist sympathizers within the Secretariat and handling the crisis in the Congo. See Robert E. Riggs and Jack C. Plano, *The United Nations, International Organization and World Politics* (Chicago: The Dorsey Press, 1988), pp. 112-119;

and Brian Urquhart, *Hammarskjold* (New York: Harper & Row, Publishers, Inc., 1972), pp. 52 and 476. Also, note Harold K. Jacobson, *Networks of Interdependence, International Organizations and the Global Political System* (New York: Alfred A. Knopf, 1979), pp. 83-87 and Clive Archer, *International Organizations* (London: Allen & Unwin, 1983), pp. 152-169.

[5]Such as the National Academy of Sciences in the United States, or the Academy of Sciences in the former Soviet Union, or the Academie Francaise in Paris, the Royal Society in the United Kingdom or the Royal Society in Sweden.

[6]G/2997/XXVII of 15 December 1972; quoted in "Institutional Profiles–United Nations Environment Programme," *International Environmental Affairs* (Winter 1989), p. 65.

[7]Ibid. This was because the consensus within the international community, represented at Stockholm, was to fund an agency that would not assume major tasks of implementation and program action. UNEP's activities were envisioned on a much smaller scale.

[8]A. Leroy Bennett, *International Organizations, Principles and Issues*, 4th ed. (Englewood Cliffs, NJ: Prentice-Hall, 1988), p. 297. See also Amos Yoder, op. cit., pp. 145-146.

[9]"Institutional Profiles-United Nations Environment Programme," *International Environmental Affairs*, op. cit., p. 65.

[10]For a list of publications in the UNEP Regional Seas Reports and Studies series up to 1986 see "GESAMP, Environmental capacity: an approach to marine pollution prevention," *UNEP Regional Seas Reports and Studies, No. 80*, UNEP, 1986, pp. 51-54. Also note "UNEP, Coastal and marine environmental problems of the United Republic of Tanzania." *UNEP Regional Seas Reports and Studies, No. 106*, UNEP, 1989.

[11]See "GEMS," (Washington, DC: United Nations Environment Programme, n.d.).

[12]*New Scientist*, 10 March 1990, p. 28.

[13]Ibid.

[14]One of the most comprehensive and thorough studies on this question of the relationship between science and politics is found in Steinar Andresen and Willy Ostreng, eds., *International Resource Management: the role of science and politics* (London and New York: Belhaven Press, 1989). Andresen and Ostreng note in their "Introduction" (p. xvii) that "[t]he main purpose of ... [the] book is to analyse the role of science in political processes of international resource management. Two key problems ... [are]: 1) the role of science in initiating political processes and international cooperation, and 2) the role of science in political decision-making. The common denominator of these two problems concerns the ability of scientists and scientific knowledge to play a part in shaping and influencing political decisions on resource management."

Oran Young in his article for this book ("Science and social institutions: lessons for international resource regimes,") raises important issues regarding the ways in which scientific advice can be manipulated. A lengthy quote from his essay conveys the highlights of some important aspects of this problem: "Even more disturbing is the exploitation of science and scientists by interest groups seeking to direct processes of regime formation to their own advantage... Politics is an adversary process in which various players or interest groups use information selectively (or even manipulatively) in order to influence others or to assemble winning coalitions. For their part, scientists are generally cautious about the inferences to be drawn from their work and chary about asserting direct cause and effect relationships as opposed to *probabilistic connections*. The result, more often than not, is a situation in which scientists are deployed in such a way as to promote the interests of key players in the political process and in which scientific knowledge claims are used selectively or out of context to strengthen the case of various interest groups. Exploitation of this sort is not confined to the political right or to the political left. Conservatives in the United States and elsewhere, for instance, routinely exploit the cautious attitude of scientists regarding cause and effect relationships in their opposition to the establishment of effective international regimes to control the long-range transport of air pollutants. For their part, those desiring to ban the consumptive use of wild animals regularly use scientific findings selectively to make the case that such activities threaten the viability of species" (emphasis added, pp. 13-14; note Alan Herscovici, *Second Nature: The Animal-Rights Controversy* (Montreal: CBC Enterprises, 1985).

[15]Interview with official at UNESCO, Paris, 11 April 1989.

[16]"Activities of the UNESCO Division of Marine Sciences," prepared for the Fourteenth Session of the Assembly of the Intergovernmental Oceanographic Commission," Paris, 17 March - 1 April 1987 (Paris: UNESCO, 1987), p. 21.

[17]*UNESCO: why the S?* (Paris: UNESCO, 1987), especially "UNESCO's special contribution," pp. 9-13.

[18]Hans Ulrich Roll, *A Focus for Ocean Research–Intergovernmental Oceanographic Commission, History, Functions, Achievements, Technical Series 20* (Paris: UNESCO, 1979), p. 5.

[19]See the pamphlet "Comar, Coastal Marine Project: (Paris: UNESCO, n.d.).

[20]*Research on coastal marine systems, Review and Recommendations for UNESCO Programme 1987-1989, UNESCO Technical Papers in marine science, No. 52* (Paris: UNESCO, 1987), p. i.

[21]"Comar, Coastal Marine Project," op. cit.

[22]Interview with official at UNESCO, Paris, 11 April 1989.

[23]"Comar, Coastal Marine Project," op. cit.

[24]Ibid.

[25]Ibid.

[26]Ibid.

[27]The success has primarily been in the relatively narrow area of the promotion of scientific knowledge (among specialists, IGOs, INGOs and scientific associations such as ICSU) regarding coastal systems and their resources. More specifically, COMAR has developed regional programs in Asia, the Pacific, Latin America, the Caribbean, Africa, several states of the Arab world and in Europe. Some of the activities undertaken by the program include studies on the following components of the coastal marine environment: mangroves, coral reefs, coastal lagoons, estuarine environments, sea-grasses and traditional knowledge and management of coastal systems (from indigenous cultures). In the specific case of a COMAR workshop held in Dakar, Senegal, for coastal marine ecosystems in Africa, the participants met to plan objectives and strategy for the COMARAF (Coastal marine ecosystems of Africa) Regional Project. This included such topics as the productivity of marine systems, hydrodynamics and coastal geology, and conservation and protection of the marine systems. See, respectively, *Research on coastal marine systems, Review and Recommendations for UNESCO Programme 1987-1989, UNESCO Technical Papers in marine science, No. 52, op. cit., pp. 1-4; and Coastal marine ecosystems of Africa, Objectives and strategy of the COMARAF Regional Project, Unesco reports in marine science, No. 48* (Paris: UNESCO, 1988), pp. 9-26.

[28]Ibid. See also *Research on coastal marine systems, Review and Recommendations for UNESCO Programme 1987-1989, UNESCO Technical Papers in marine science, No. 52*, op. cit., (emphasis added) p. i.

[29]*Report of the Director-General on the activities of the Organization in 1984-1985 communicated to Member States and the Executive Board in accordance with Article VI.3.b of the Constitution, 24 C/3* (Paris: UNESCO, 1986), p. 47.

[30]Interview with official at Intergovernmental Oceanographic Commission, Paris, 11 April 1989. See also Hans Ulrich Roll, op. cit., pp. 31-33.

[31]Hans Ulrich Roll, op. cit., pp. 5-10.

[32]Ibid., p. 26.

[33]Ibid.

[34]Ibid., pp. 26-27.

[35]See Edward D. Goldberg, *The Health of the Oceans* (Paris: The UNESCO Press, 1976).

[36]This paragraph is drawn from Hans Ulrich Roll, op. cit., p. 33.

[37]Ibid., p. 36.

[38]Pamphlet, "IGOSS, The Integrated Global Ocean Station System," (Intergovernmental Oceanographic Commission and World Meteorological Organization, 1979), p. 3.

[39]Hans Ulrich Roll, op. cit., p. 36.

[40]Ibid., p. 41.

[41]This section is drawn from Ibid., pp. 42-43.

[42]Ibid., p. 31.

[43]This paragraph is drawn from Ibid.

[44]Ibid., p. 32. The rest of the seven areas are: "participation in oceanographic cruises; assessment of training needs by convening small groups of experts on a regional basis; use of ... [an] IOC [t]rust [f]und for TEMA purposes; annotated bibliography of marine science textbooks, reference books and manuals; directories of experts in the marine sciences; participation of young foreign scientists in research and training projects of developing countries; and [lastly a] programme of work and budget for TEMA." The last is certainly a useful and sometimes overlooked component to the international bureaucratic existence.

[45]See *Marine Science and Ocean Services for Development: UNESCO/IOC Comprehensive Plan for a Major Assistance Programme to Enhance the Marine Science Capabilities of Developing Countries* (Paris: UNESCO, 1985), p. 24.

[46]This section is drawn from the descriptive inside back cover of the *WMO Hydrology and Water Resources Programme 1988-1997, Second WMO Long-Term Plan, Part II, Volume 5* (Geneva: World Meteorological Organization, 1987).

[47]Ibid., p. 5.

[48]This section is drawn from the descriptive inside back cover of the *WMO Hydrology and Water Resources Programme 1988-1997, Second WMO Long-Term Plan, Part II, Volume 5* (Geneva: World Meteorological Organization, 1987).

[49]The six remaining major scientific and technical programs are: "1) *The World Weather Watch Programme...* It combines data-processing centres, observing systems and telecommunication facilities–operated by Members–to make available meteorological and related geophysical information that is needed in order to provide efficient meteorological and hydrological services within the countries... 2) *The WMO Research and Development Programme* promotes atmospheric research, placing highest priority on weather-prediction research on all time scales... 3) *The WMO Applications of Meteorology Programme* comprises three ... areas of application of meteorological services and information: agricultural meteorology, aeronautical meteorology and marine meteorology, and promotes the development of infrastructures and services which are required in all three areas for the benefit of ... Member countries. 4) *The WMO Hydrology and Water Resources Programme* is concerned with the quantitative and qualitative assessments and forecasts of water resources; standardization of all aspects of hydrological observations; and the organized transfer of hydrological techniques and methodology in many areas, including forecasting mitigation of floods related to tropical cyclones, severe storms and rapid snowmelts. 5) *The WMO Education and Training Programme* ... [promotes] efforts in Member countries to ensure that the necessary body of trained meteorologists, hydrologists, engineers and technicians is available. [And] 6) *The WMO Technical Co-operation programme* comprises the mainstream of

organized transfer of meteorological and hydrological knowledge and proven methodology among ... Members... Particular emphasis is laid upon the development of a wide range of services (related to weather prediction, climatology and hydrology); on the development and operation of key World Weather Watch Infrastructures; and on supporting the Education and Training Programme of WMO." See inside back cover of *The World Climate Programme, 1988-1997, Second WMO Long-Term Plan, Part II, Volume 2* (Geneva: Secretariat of the World Meteorological Organization, 1987).

[50]See *World Ocean Circulation Experiment Implementation Plan, Vol. 1, Detailed Requirements* (Wormley, England: WOCE Planning Office, 1988), inside front cover.

[51]Ibid., from Forward.

[52]Ibid. G. A. McBean served as chair of the Joint Scientific Committee for the WCRP (World Climate Research Program) and A. D. McEwan served as chair of the Committee of Climate Changes and the Oceans. There are three "streams" or time periods of research involved in the WCRP. The first relates to understanding weather patterns of one to two months. The second concerns "... global climate over periods from several months to several years..." And, finally, the third "... aims at characterizing the variations of the atmospheric climate over periods of several decades and assessing the potential response of climate to either natural or ... [human] influences, such as the increase of the concentration of atmospheric carbon dioxide." These efforts are coordinated jointly by the WMO and ICSU. See *The World Climate Programme 1988-1997, Second WMO Long-Term Plan, Part II, Volume 2*, op. cit., pp. 52-53.

[53]Ibid.

[54]Ibid.

[55]*World Ocean Circulation Experiment Implementation Plan, Vol. II, Scientific Background* (Wormley, England: WOCE International Planning Office, 1988), Section 3.1.1.

[56]Atmospheric forcing is a long-term average of weather patterns formed by such climate considerations as wind and weather fronts. These climate factors are related to surface currents in the ocean. I am indebted to the U.S. National Weather Service in Slidell, LA for this explanation.

[57]The budgets for these programs are not in the billions of dollars (US). Far from it. The primary funding falls upon national governments, IGOs, INGOs and private agencies participating in the various projects. It is "big science," though, in the sense of being *very large programs* with ambitious goals. In this sense, it is similar to the "big science" characterized by such projects as the Space Station *Freedom*, the Human Genome Project, Cassini Saturn Probe or Mars Observer, or the Superconducting Supercollider. For a fascinating description of these programs,

their projected costs and impact on science and science funding in the United States, see *The New York Times*, 27 May 1990, p. 1 and 10 June 1990, p. 1.

[58]See the Preface by D. W. H. Walton in D. W. H. Walton, ed., *Antarctic Science* (Cambridge: Cambridge University Press, 1987), p. vii.

[59]Ibid. Also, see Philip W. Quigg, *A Pole Apart, The Emerging Issue of Antarctica* (New York: McGraw-Hill Book Co., 1983), p. 2.

[60]This paragraph is drawn from *World Ocean Circulation Experiment, Implementation Plan, Vol. II, Scientific Background,* op. cit., Section 4.1.

[61]Ibid., Section 4.1.1.

[62]This paragraph and the next are drawn from Ibid., Section 5.1.

5

INGOs and Cetacean Protection

Introduction

INGOs play a distinctive role at two levels. The first is at the international plane, and the second is at the level of the individual nation-state. This chapter will begin first by examining the unique and growing impact INGOs have in shaping new and original ideas that heretofore have not been prominent in the global political debate. Then the impact of INGOs on decision-making in the context of the nation-state itself will be examined. In particular the work and activities of two INGOs will be analyzed: The Cousteau Society, and the international environmental group collectively known as Greenpeace. The chapter will close with some brief concluding remarks on the nature of INGOs within the international system.[1]

INGOs have grown considerably in numbers since the end of the Second World War. It is, therefore, interesting to see what this development has meant to the political problems associated with the global ocean. In addition, INGOs are now participating in the political debate at the level of the nation-state to a degree greater than at any other time, and they also are participating in global conferences as a matter of routine. This has had significant repercussions for the final agreements reached through such multilateral fora. Finally, and most importantly, INGOs are often catalysts in developing new and original thinking on common problems confronting the whole of the international community. Co-existing with and operating alongside national governments in many areas of common concern, but neither subservient to nor in agreement with them at all times; INGOs raise serious issues and problems, as well as propose solutions not mirrored, for instance, in the policy choices of national governments. The same can be said for national delegations sent to international conferences. These points will be explored in the pages to follow.

Ideas and the Public

INGOs, which have proliferated since the First World War, are a unique and permanent manifestation of contemporary international society. However, INGOs operate in a realm which is not only different and somewhat removed from the concerns and pressures of national governments and IGOs; they are also part of the semi-independent international structure that co-exists in tandem with the inter-governmental nature of the contemporary state system.

These "transnational" organizations have been described in the following way:

> The INGOs carry out a variety of border-crossing activities to attain their goals in the pursuit of interests for which they have been created. These activities create relationships of the INGOs with both governmental and non-governmental entities and actors; such relations have been labeled trans-national in contrast to traditional international relations, which are generally understood to apply only to activities and contracts between governmental actors.[2]

By operating somewhat outside "traditional international relations," INGOs have been able to criticize national governments and IGOs or to propose solutions which can serve as alternatives to existing national or inter-governmental programs or policies. It is axiomatic to note that if the INGOs wish to be effective in achieving or advancing the goals for which they were formed, they must reflect (at least in a broad way) the realities of the international system.

Nonetheless, INGOs have had a growing and significant impact on the actions of both governments and IGOs. An example is The Club of Rome, one of the first to gain widespread and meaningful influence.[3] Alexander King, a co-founder of the organization, has written:

> The central concern of the Club ... is with what it calls "Le World Problematique," that tangle of interacting problems which we now face and within which it is increasingly difficult to formulate discrete issues and discrete solutions. The perspective ... is thus essentially global. It has two functions: 1) Stimulation of research ... [and] 2) Bringing the results of its findings to the attention of decision-makers in government, industry, and so forth, and holding dialogues with such people concerning the increasing importance of global problems and of longer-term thinking. In fulfilling these functions, the Club restricts its activities to those of a catalytic nature.[4]

Publications sponsored by The Club of Rome have helped to raise the level of awareness with regard to common global problems. The functions referred to by King in the preceding quotation found expression in two studies of the early 1970s which have influenced public opinion and governments in the West, particularly in the wake of the Oil Crisis in October 1973 (following the fourth Arab-Israeli War). *The Limits to Growth* and *Mankind at the Turning Point*, contributed to an appreciation of the predicted dire straits into which humanity appeared to be headed if then current trends were continued.[5]

The Club of Rome has undertaken a variety of studies on global structural problems, besides the original work involving economic and demographic constraints on growth. Alexander King has also written that while "the initial studies of the Club ... concentrated mainly on the material aspects of the world Problematique ...," great attention has also been paid to "social and political aspects ..." that significantly impact on the "value system" inherent in Western culture. Club sponsorship of such expanded topics has included 1) "goals for harmonious survival of the various societies of the world," 2) a study of "world power structures [both] formal and informal," including state and non-state factors; and 3) "educational requirements for a new world [order]." Ironically, the last project was put forward at the suggestion of Romania, during the brutal Ceausescu dictatorship.

Through its notoriety and access to decision-makers, at both the national and IGO levels, and because of the respect it has generated, The Club of Rome has exercised significant influence in the ongoing thirty-year debate over various preferred global futures. "It is difficult to state categorically how deep the political influence ... has gone," however.[6] This can be said about other INGOs as well. A 1988 study noted the increasing importance of INGOs within the international system, yet also raised the difficult question whether such influence "has resulted in significant shifts in governmental or IGO policymaking..."[7] The study attempted to answer this question by looking at two important issue areas: human rights and disarmament.[8] The subject of this chapter will focus, however, on the issue of ocean degradation and the protection of cetaceans, specifically dolphins.

At this point in time, marine biologists and other specialists do not know what the effect would be if dolphins were to become an extinct life form. Is it just a matter of aesthetics? Should human beings be concerned about these species only because they are beautiful and inherently have a right to exist; or, is the dolphin somehow intimately related to, say, the successful operation of the food chain in the global ocean? If the latter is true, then one can argue that the matter assumes much greater urgency within the plethora of fundamental challenges facing *humankind* in the last years of the 20th century.[9]

International Scientific and Professional Associations

International Scientific and Professional Associations (ISPAs) in general have significantly contributed to the enhancement of environmental concern and action. While The Club of Rome is, of course, a discrete entity, ISPAs comprise many formal groups or affiliations.

> The ISPAs range from relatively loosely structured associations (such as the Pugwash group, which has held over twenty-four conferences and twenty-five symposia, spread out over twenty-five years) to complex associations such as the Committee on Space Research (COSPAR) ... ISPAs are forums for organizing and maintaining international contact among their membership, particularly with a view to the presentation and stimulation of research and the diffusion of findings.

In addition, "they are forums for the planning and organization of cooperative international research projects such as the International Geophysical Year (IGY)" undertaken in 1957. As Feld and Jordan also note, these international organizations "are ... forums in which compromise solutions to seemingly intractable international problems can be explored, and in which mutual understanding can be developed."[10] ISPAs are ubiquitous. One estimate is that there were a little over 900 present in the 1980s alone.[11] One can reasonably assume the numbers have since increased.

Those scientists, leaders in business, other academics, including members of government and governmental bureaucracies that are members of ISPAs, can influence decision-making in several ways.[12] First, at the level of the nation-state, members can lobby national governments in opposition to, or support of, specific programs or policies. They can also serve as bridge-builders, so to speak, between various or disparate interests. Members frequently perform the function of expert witnesses or available specialists ready to consult with relevant decision-makers. They often serve as members of special committees or commissions set up to study particular issues of a technical or specialized nature. When members travel to international meetings for their respective governments, i.e. as participants at an intergovernmental governmental organization (IGO) assembly or conference, they are often utilized as expert informants or counselors. They can wield influence in that forum by making use of a private word or discussion with appropriate delegates or authorities, or their intervention can be more public and open if they are delegates themselves with the right to make statements in general plenums. Finally, at the international level, members of ISPAs can influence policy as "technical advisors or consultants to international secretariats, as lobbyists, or as members of international secretariats." In the latter category "they can have an influence on the articulation and implementation of internationally agreed policies."

Diana Crane has written of an "invisible college" created when experts from technical, scientific and social science disciplines interact at the level of an international "informal network." This network is characteristic of much of the day-to-day work of technical specialists and academics when they are not gathered together for formal meetings or conferences. Crane defines this "invisible college" as "the informal network of experts who are in contact with one another but who do not share a single organizational affiliation. Such information groups, rather than the ISPAs, may actually be the most important actors in such situations [of international consultation or interaction]."[13] This "invisible college" is very similar to "networking," a term that has come to mean the cultivation or association of like-minded professionals seeking to influence, develop and share useful information, ideas and contacts that may prove to be mutually beneficial to the individual or the organization, or both.

The importance of the opening of "new channels for communication, information exchange, and expert evaluation ..." which ISPAs provide should be noted.[14] It is axiomatic to state that the global telecommunications media conveys the debates, questions, controversies, and successful discoveries, as well as the set-backs and failures, of the international research and academic community (this includes especially the work of those in the physical and natural sciences). By their participation in the praxis of research, as well as their involvement in the process of advising national governments and IGOs, the members of ISPAs help shape decision-making, debate and discourse.

INGO Influence on National Decision-Making

The Cousteau Society

The activities of INGOs can overlap at the level of the nation-state and the intergovernmental organization (IGO). In other words, many INGOs engage in activities of public information, lobbying, testifying before authoritative bodies or committees, or participating as official delegates or observers at international conferences at two stratums: the domestic and IGO tiers. This is not always the case, but many times it is so. These INGOs participate in political activities which straddle the common distinctions usually made between domestic and international politics. For example, The Cousteau Society (a membership organization made up primarily of American and French citizens, but not confined to those nationalities[15]) expends much of its organizational efforts generating, focusing and enabling public opinion to center on issues which the society considers to be of extreme importance, i.e. preservation of the biodiversity of the Antarctic. (This theme will be resumed later in the chapter.) The Cousteau Society also, however, does not limit itself only to "global matters." Particularly in the area of offshore oil leasing by the United States government, officers of the society (especially in recent years Jean-Michel Cousteau) have made strong efforts to make known the position of the organization. For example, testifying before U.S. Congressional committees on the extension or modification in the scope of oil drilling activity and platforms off the California coast is an area of long-standing society interest and concern.[16]

The Cousteau Society does not engage directly its political adversaries, nor does it participate in the more straightforward lobbying activities normally associated with political parties or pressure groups, i.e. running candidates for public office, engaging in advertising that directly attacks individuals or groups for their political views, or sponsoring such activities by others.[17] It *does* engage in politics, though, in the sense of attempting to put across to political leaders and the general public its' views on questions of public policy. And, it also "lobbies" in the sense of participating in educational events—such as lectures and vacation/study tours—which seek to raise the level of public knowledge or awareness with regard to issues of the integrity of the ocean. The primary activities of the society include production of the now famous television programs, lectures before public audiences, testimony before congressional and parliamentary committees or commissions, and what can be called the sophisticated and calculated utilization of the "Cousteau Interview" or the "Cousteau Appeal" to elected heads of government. This latter unique and finely crafted instrument has become a carefully developed tool of the organization's founder, Jacques-Yves Cousteau, and his associates. There is no other INGO in existence today that has at its disposal such an individual media-agent with access to high government officials.[18] In short, the primary focus of The Cousteau Society is on public education.[19]

The activities of the Society in mobilizing world public opinion to focus on environmental threats is perhaps nowhere more profound, and ironic, than in its' identification and coverage of past ecological disasters.[20] It is profound because of the power of television, and ironic because environmental groups still seem to be more successful in explaining why and how past events occurred, than in preventing similar happenings from occurring in the future. Of course, the changes that environmental INGOs—operating in industrial market democracies—are attempting to bring about are ones that take time. In that sense, they are "developmental." And, because of the structure of the international system and the constraining circumstances of not having the resources at their disposal to the extent which national governments do; it is unavoidable that INGOs oftentimes suffer in head-to-head confrontations with their erstwhile IGO "systemic partners."

The *Exxon Valdez* is a case in point (see Chapter 3). Within hours of news of the wreck in the Prince William Sound area Jean-Michel Cousteau began gathering a film team and flew to Alaska. There he and society photographers recorded images of the destruction and the spill's impact, as known at that time, and interviewed participants in the drama (including oil company officials, Coast Guard representatives, local residents and those engaged in the fishing industry). A year later Cousteau went back to the shores of the bay and produced an hour long television program detailing, as he saw it, the continued and likely long standing damaging environmental impact caused by the spill. The society, and the elder the Cousteau in particular, have been criticized in the past for the research methodologies utilized (or the lack thereof). The criticism boils down essentially to this: A lot of the work is simply bad science. It is charged the work is not detailed enough in the complexities of the subject being analyzed; and, most importantly of all, the whole operation tends to "showmanship" which operates at cross purposes with the scientific ethos.[21]

It is not the task here to analyze the scientific contributions of Jacques-Yves Cousteau and his very many collaborators since the end of the Second World War. As an INGO, however, The Cousteau Society has participated in numerous international exercises or undertakings which have directly sought to influence decision-making at the level of the nation-state. In the Society's history such actions have focused primarily, but not exclusively, on the United States, and more recently with the Antarctic and Wellington Treaty issue the organization has broadened its efforts in "lobbying" at the nation-state level.[22]

The links between domestic public opinion groups and the activities of INGOs has received prominent attention. Feld and Jordan, as well as Lynton K. Caldwell, have investigated these relationships. The former have found "that ... INGOs are recognizing that links directly to domestic public opinion play a role in the formation of governmental policies... [B]y relying on this direct access, and by carefully concentrating on issues that would be likely to provoke public response, INGOs hope to pressure their governments more effectively."[23] In other words, purposeful intervention is a frequently utilized strategy of INGOs to interject themselves into national political debate. And, Caldwell has noted (perhaps perspicaciously) the "formal" and "informal" components to this intervention.

> Formal or informal interventions by ... [INGOs] with political and administrative authorities are the more frequent method of ... [INGO] participation in the implementation process. But where governmental action appears to ignore or contravene international obligations as perceived by ... [INGOs] stronger measures may be taken. How strong and in what form depends upon the tolerance of a national government or judicial system for opposition and the degree of ... [INGO] dedication...[24]

Caldwell's observation on the implementation components to the work of INGOs holds true when the focus of analysis is shifted to groups such as the one being discussed here.

> In efforts to implement international commitments the role of ... [INGOs] has become a regular and anticipated part of the action. All of the larger environmental ... [INGOs] with international missions monitor the administration of treaties and other international agreements. They report regularly to their memberships and to elected officials on observed

> performance, and may identify failures or insufficiencies in program implementation.[25]

This needs little elaboration. By these types of actions, groups such as The Cousteau Society participate in a significant way in the global political debate regarding alternative and proffered environmental futures.

Greenpeace

Greenpeace is another INGO which has purposely sought to influence the policy process at the domestic nation-state level, as well as pursue its political objectives at the IGO level. Greenpeace, with 4 million financial supporters planetwide, has offices in many Western industrial countries; including the United States, Canada, the United Kingdom, the Netherlands, Belgium, Australia, Norway, Spain, Sweden, Switzerland and Germany.[26] These semi-independent, yet interconnected, national "environmental action committees," through their membership and professional staff, represent a well-organized, committed and very active "soldiery" which provide the various national organizations with a ready support and means of execution for agreed upon political strategies. It goes without saying, of course, that these varied domestic sources provide the national Greenpeace organizations with the personnel and funds for their environmental programs and activities (including the international component when they choose this to be an appropriate venue).

The self-appointed Greenpeace environmental brief is wide-ranging. A media packet released in the Fall of 1989 states clearly the motivating vision as well as the pragmatic goals and objectives.

> Greenpeace is an international organisation dedicated to the protection of … [the] natural environment … [and] campaigns to protect the environment from nuclear and toxic pollution, to halt the slaughter of species such as whales and seals, to stop the testing and production of nuclear weapons and to curtail the arms race at sea. Greenpeace also campaigns against the mining and reprocessing of nuclear fuel, the exploitation of the Antarctic continent and the destruction of marine resources through indiscriminate fishing methods.[27]

This section will focus on the specific Greenpeace programs regarding dolphins, as these activities are primarily made manifest through political action within the United States and the London Dumping Convention, the latter a treaty entrusted to the International Maritime Organization (IMO), headquartered in

London. By focusing specifically on the issue of cetacean protection, the impact Greenpeace on the policy processes both at the domestic or national and international levels can be revealed.

It is helpful before beginning an examination of these issues, to note what animates the Greenpeace philosophy: "Native American prophesy tells us that when the Earth is sick and dying, all over the world people will rise up as Warriors of the Rainbow to save the planet. This legend inspired Greenpeace to call its flagship the Rainbow Warrior."[28]

The message of impending environmental disaster is well supported by the empirical investigations being undertaken in a variety of scientific disciplines.[29] Greenpeace's commitment to "direct action" however, is more contentious and controversial.[30] Thus, the organization attracts wide attention both nationally and internationally. The professional staff is also growing in size, as is the sophistication in methods employed to reach their policy objectives.[31]

Greenpeace USA (the official name of the Washington, DC based group)[32] has been deeply involved in the conservation and cetacean protection issue in the Eastern Tropical Pacific because of the role historically played in this area by U.S. registered fishing vessels and because of the size of the domestic U.S. canned tuna market.

For example, Greenpeace has been very active in lobbying before the U.S. Congress seeking legislation that would ban purse seine nets which are used to catch tuna for commercial fishing operations. This issue has assumed a great importance because through this type of fishing dolphins are taken in large numbers and many die as a consequence. American-flag ships are only a small portion of the total tuna fishing fleet, however.

Dolphins, for reasons not yet known, associate in herds with numerous species of tuna, including skipjack and yellowfin (two of the most important ones sought for canning operations, and hence for retail sales in the industrial nations; especially the United States for reasons detailed below). The dolphins swim above the tuna while moving through the open ocean, and when the fishing boats locate dolphin herds the current operative technology is to lay out a large purse seine net which surrounds the exploitable commercial species swimming underneath. Unfortunately for the dolphins they are also caught in this operation.

The issue is an exceedingly complex one not only because of economic considerations (those engaged in the commercial fishing industry, and consumers), but more pointedly the technology employed along with the changes that have been required by national governments–particularly the United States, which is the largest market for canned tuna in the world–has evolved markedly over the past several decades.[33]

The modern or contemporary tuna industry, in the Eastern Tropical Pacific, developed in the 1920s and as recently as thirty-five years ago "the majority of tuna was caught on hooks from bait boats and longliners." As Lesley Scheele and Dean Wilkinson have noted: "Such vessels were small and usually fished within 200 miles of the mainland or in the vicinity of offshore islands such as the Galapagos. Bait boats dominated the industry with 80 percent of the total capacity. The few purse seiners which did exist were small, with capacities of 200 tons or less, and of minor importance to the fishery overall."

The Scheele and Wilkinson paper is an extensive analysis of the issues related to the tuna industry operating in the Eastern Tropical Pacific, including analysis of fleet composition, the Marine Mammal Protection Act of 1972 (passed by the U.S. Congress and signed into law by President Richard Nixon), amendments, permits and regulations associated with the 1972 act, and performance standards and parameters of industry operation. It is not necessary to raise all these issues in order to establish that the activities of Greenpeace regarding cetacean protection and conservation spring from the advances in technology which have made the problem of how tuna is commercially caught such a critical issue in environmental protection.

By 1958 and 1959 certain developments within the tuna fishing industry became quite strong. Most important among these were the creation of a "puretic hydraulic power block which allowed for the rapid recovery of much larger nets." Lack of such a capability had previously hampered fishing boats by decreasing net size and by slowing the capability of the boat to quickly retrieve the net. Larger nets were also created ("3/4 of a mile long and over 300 feet deep") to efficaciously encircle the dolphin herds. Fast-powered speed boats were also brought into the industry, and these were used to chase and confine the dolphins (and tuna swimming below them) while deck-hands on the fishing boat laid out the line. As Scheele and Wilkinson note, "The result of these changes was record landings of valuable yellowfin and massive dolphin mortalities."[34] The Inter-American Tropical Tuna Commission (headquartered at Scripps Institution of Oceanography in La Jolla, California and made up of national members from the United States, Japan, France, Panama and Nicaragua) in 1981 estimated that in 1959 approximately 590 "sets" or convergences of persons and technology were made on various dolphin species. In 1960 the number dramatically increased to 5,400 "sets."[35] Technological improvement has not, of course, been in abeyance since the early 1960s. However, the basic parameters of the problem have, according to Scheele and Wilkinson, remained essentially the same. Their lengthy observation on this point is illuminating.

> The procedures used to set on dolphins are much the same today as when the fishery first began to exploit the associative bond. When a herd of dolphins is spotted, three or more speed boats are launched from the seiner. The speed boats travel behind the seiner until the larger vessel is considered

> close enough to the herd. The ... boats then race ahead and attempt to surround the dolphins and stop their forward movement. By this time the dolphins have begun to run–an avoidance behavior that rarely works to their advantage because their movement is linear, predictable, and very exposed. Wakes from the speed boats and the boats themselves are utilized to compact the herd and bring it to a stop. This may continue for up to 90 minutes. When the herd is stationery, net deployment begins.[36]

By this point the dolphins are apparently very tired. They do not jump over the nets when the operation has successfully encircled them and the tuna, and very often the dolphins suffer anatomical damage to their flippers, rostrums (beaks) or flukes. As air-breathing animals, dolphins will suffocate if they become entangled in the nets. They must surface in order to breathe.

It should not be thought, however, that those involved in the fishing industry have no concern at all over the threat to dolphin species. A rational economic self-interest operates: dead dolphins are of no future use in locating tuna. Certain techniques were instituted on the tuna boats which have resulted in a decrease in dolphin mortality.[37] The chief ingredient to this development was the "backdown" policy. This really quite simple and unsophisticated procedure has had significant results.

> [I]t is still the single most significant contributor to mortality reduction. Backdown begins when 1/2 to 2/3 of the net has been retrieved. The vessel reverses direction, causing the net to elongate into a narrow channel. Water flow sinks the corkline at the far end allowing dolphins to pass over the net. Manned rafts, stationed in the backdown area, further assist the dolphins to escape. If the tuna attempt to follow, the vessel is stopped, and the corkline is raised.[38]

This has not provided an acceptable answer to the problem, however, as will be discussed below.

A useful base line for the tuna industry can be established in the year 1960. At that time, U.S. tuna boats accounted for 99% of all tuna boats operating in the Eastern Tropical Pacific (ETP).[39] By 1986, of those boats using purse seine netting (and for all practical purposes the entire industry made this switch approximately more than two decades previously), only 33% "... of the purse seiners setting on dolphins in the ETP were of U.S. registry."[40]

There are several reasons for the diminishing numbers of the overall U.S. composition in fleet totals. Certainly the entrance of foreign owned, operated and financed boats has played a significant part. In addition, however, U.S. vessels have transferred their operations to foreign flags in order to escape the increased costs of doing business in the United States (as compared, say, with most Third World and Newly Industrialized countries); and they have transferred their operations attempting to escape from the spectre of increasing regulatory intrusion from governmental bodies.[41] From the point of view of the tuna industry, activities of groups such as Greenpeace compound their economic and management problems.

Greenpeace has actively lobbied for the extension and enlargement of the provisions of the Marine Mammal Protection Act of 1972. This original approach in conservation, management technique and philosophy was, when first passed, a unique departure in policy making for the U.S. Congress.

> Although the legislative process had attempted to accommodate disparate views ranging from doing nothing to complete protection, what emerged was a milestone in wildlife legislation which went far beyond the borders of the United States in its impact... [I]t signaled a strong U.S. commitment to preserving marine mammal populations.
>
> The Act placed an immediate moratorium on the taking and importation of marine mammals and marine mammal products. At the same time, it allowed exceptions so long as they were consistent with the goal of protecting marine mammals. Prohibitions and restrictions were placed on the intentional kill, incidental kill, and even harassment of marine mammals.[42]

As a product of the legislative process and the pluralistic political culture within the United States, the act is a flawed instrument from the point of view of outright or unqualified protection of dolphins and other marine mammals. Even the maximalist proponents of the legislation and the policy advocates supporting the position from the INGO realm understood this from the very beginning. The first legislative step ever taken in the United States in the direction of expanding human concern for the protection of other mammals was not one of unqualified or complete defense. Greenpeace, however, has continued to lobby for full implementation of the law—which has not occurred in the many years since its passage—and for strengthening the objectives of the original legislation itself.[43]

Particularly throughout the 1980s Greenpeace, along with numerous other environmental groups active within the United States, aggressively lobbied for measures to enhance and extend the provisions of the original act.[44] One area which has garnered significant attention from the environmental community is the importation of tuna from other nations into the American market. Those nations have generally had weaker regulations covering the killing of dolphins in tuna fishing operations. Greenpeace has charged that the relevant executive bureaucracies within the United States Government—specifically the National

Marine Fisheries Service (NMFS) of the Department of Commerce—have not been committed to implementing the legislation passed by Congress. The legislative branch passed amendments to the original act in 1984 which included prohibitions on the importation into the United States of tuna from other nations "... whose kill rates and regulatory program were not comparable to those of the U.S. fleet."[45] Greenpeace has been continually at odds with the Commerce Department regarding provisions of the act, especially the way in which the NMFS has administered the law.[46]

The Scheele and Wilkinson study, one of the most comprehensive research efforts undertaken on this issue released for public dissemination, although only approximately 60-pages, is replete with simple and heuristically useful statistical tables illustrating the declining number of dolphins in the study area almost twenty years after legislative action was taken ostensibly to prevent exactly what has transpired. Utilizing a variety of sources[47] the authors give strong evidence to demonstrate that over 6.2 million dolphins died from 1959 through 1989 in the Eastern Tropical Pacific alone. This extremely conservative estimate is reached by accounting only for those dolphins found dead in the nets of the tuna fishing boats. Those that died after being released, with broken beaks or flukes, for example, are not counted.[48]

Because implementation of the legal, administrative and legislative functions has proved so difficult from the point of view of the objectives of Greenpeace, the organization broadened its approach to the problem of dolphin protection by embracing additional strategies. While previously focusing lobbying and scientific education efforts on Capitol Hill and in the offices of the various executive departments, and at times through the courts, Greenpeace decided to widen its "playing field" by attempting to motivate its membership, other sympathetic groups and the general public at large to boycott the tuna canning corporations that caught and sold tuna in the usual manner, i.e. from suppliers using purse seine netting. As the reader knows, a probable symbiotic relationship exists between certain species of tuna and dolphins.

In 1989 legislation was introduced in the House of Representatives to label tuna sold in the United States, regardless of the location where it was caught or under whatever flag the ship was operating. Both Representatives Barbara Boxer and Mel Levine sponsored the legislation (H.R. 2926 and 2948, respectively), which were very similar in content. Specifically, the "... Boxer Bill would require that all tuna products destined for [the] U.S. market must be labeled to indicate whether or not the product was caught by a method known to kill dolphins."[49] Greenpeace combined its support of this legislation with an attempt to achieve through market pressures what was not possible through the regulatory format.[50] By sponsoring, along with other organizations, a consumer boycott of tuna products, Greenpeace sought to use economic pressure against the large canners and distributors to reduce the incidence of death among dolphin species.

The largest tuna canner in the world is the American company Star-Kist Seafood, a subsidiary of H.J. Heinz Company. In April 1990 Heinz announced "... it would no longer buy tuna trapped by nets that kill dolphins."[51] For the environmental groups this was a major breakthrough and ended a tortuous three year battle which pitted Greenpeace, and other organizations, against the largest American corporations purchasing, manufacturing and marketing canned tuna products.[52] Interestingly enough, however, Starkist's revenue and profit "... actually grew during the two years of the boycott called by..." the environmental activists.[53] This is not the place in which to examine the complexities of the American market. It does raise the inviting question why did the American canning industry decide to comply with the demands of the boycott if profits and revenues actually rose during the period 1988-1989. This venue, too, is not an appropriate one in which to deeply delve into this question. One can reasonably posit, however, that the final decision had much to do with public relations, the changing contours of environmental concern as understood by American public opinion, and the desire to lay to rest a contentious issue that occupied a disproportionate amount of time on the part of senior management. Greenpeace applauded the action, but stated verification procedures must be implemented to ensure that "dolphin safe" tuna is the only type of tuna available on store shelves in the United States.[54]

Another overlap between INGO domestic and international political activity concerns the issue of driftnetting. Greenpeace has played a prominent role in opposing this fishing technique and in efforts to eliminate the practice altogether. They have vigorously, through a variety of forums, criticized nations that utilize driftnetting methods, including the lack of perceived responsiveness of the United States Government in supporting measures to negotiate the ending of this practice by the three main Asian nations that engage in it—Japan, the Republic of China (Taiwan) and the Republic of Korea (South Korea)—and have assiduously lobbied members of the U.S. Congress to support their policy objectives.[55]

The driftnet is a sort of "gill net" of woven monofilament fishing line. Each square in this net is approximately 3 inches in diameter, and is extremely effective in catching any fish or mammal that swims or flies into it. When a life form becomes entangled in the net it is very difficult, usually impossible, for it to escape. The results are devastating. The technology of this industry has expanded so that driftnets can be 15 to 30 miles long and 25 to 40 feet deep. Weights are placed at the top and bottom of the nets to give stability and are laid-out in the water in an "array-like" configuration. It has been estimated by Greenpeace and other organizations that during each nocturnal fishing period enough monofilament line is laid to cover many thousands of miles.[56] Driftnetting techniques and practices have been blamed for deleterious effects on fishing stocks as well as on cetacean mammals (dolphins, porpoises and other odontocetes, i.e. toothed whales).[57]

> Driftnets are greedy. They destroy everything they touch—not just the target fish, like squid, but also salmon, whales, dolphins, seals and sea birds. And nearly half of all marine life, including the target fish, falls out before the net is hauled in. They're cheap and easy to use, and because they're made of lightweight, nearly invisible plastic threads, fishermen can play out ... miles of net at once.[58]

Driftnetting is the logical extension of the Industrial Revolution ethos *ad infinitum*. It is an extreme manifestation in the global ocean today of the uncontrollable, short-sighted and exploitative nature of humanity's relationship with the environment. It is symptomatic of the severed human connection that has developed during approximately the past two and a half centuries between what is sustainable in human life and what is not.

The June 1990 Greenpeace film showed to interested Members of the United States Congress highlighted specific impacts on marine life of driftnetting in the Tasman Sea (a body of water roughly between Australia and New Zealand). The main Greenpeace research vessel—the Rainbow Warrior—left for the area in January 1990 with a team of scuba divers, videographers and a marine biologist.[59] What they found was graphic.

Greenpeace estimated a conservative mortality figure for dolphins during the three month long fishing season in the area of 6,000 deaths. They also claim the video program is the first documentary evidence of "this wall of death." Roger Grace, another marine biologist involved in the project (but not aboard the ship during the actual expedition), described the technique as "very unmanageable and wasteful." Many fish caught in the nets are ones for which no market exists in which to sell the product. Examples abound. Industry sources acknowledge many unsuitable and unusable fish are caught. Greenpeace estimates during the season under investigation over 3,000 sun fish were killed in driftnets. Besides the dolphins, over 4,000 sharks were trapped and killed, as well as 3,000 bill fish (marlin, sailfish, et al.). This type of massive environmental impact raises serious questions regarding the viability of the food chain in the Tasman Sea. The simple numbers alone, of the different species taken, without reference to any other variables, is clearly a serious indicator of environmental threat. It also goes without further emphasis that if the Greenpeace figures are broadly accurate (and there is no evidence to suggest they are not) the impact in the Tasman Sea, and perhaps beyond, are just the effect of one season's fishery. Multiplied by several years the reader can well imagine the disastrous repercussions on the local environment. Finally, because driftnets are not biodegradable, when the nets are cut to dispose of unwanted or dead fish the remnants falling into the water contribute to plastic pollution. Grace described the dolphin mortality figure as a "very dangerous precedent."[60] The pillage and plunder intrinsic with this technology is patently ominous.

Because of their concern for the implications on the viability of marine life, Greenpeace has actively opposed the continuation of this fishing method. Through videofilms as discussed above, and through lobbying efforts at the United Nations headquarters in New York, the organization has sought a blanket and "... immediate halt to [all] driftnetting..."[61] With offices in forty nation-states, these semi-autonomous Greenpeace organizations recently decided to send to the United Nations representatives from the various Greenpeace member countries to lobby their own national missions. Greenpeace USA has noted that the pursuit of this route, to influence member-states of the United Nations organization, will help to avoid criticisms that the campaign is primarily an American initiative.[62]

As a sign of what may appear to some as a success, in December 1989 the General Assembly adopted a resolution addressing the issue of driftnet fishing in the South Pacific and other areas of the global ocean.[63] The resolution is weak, however. The three relevant paragraphs of the resolution are as follows and were offered as a "recommendation" to the relevant nation-states.

> a) Moratoria on all large-scale pelagic driftnet fishing on the high seas by 30 June 1992 *with the understanding that such a measure will not be imposed on a region, or if implemented, can be lifted, should effective conservation and management measures be taken based upon statistically sound analysis to be jointly made by concerned parties of the international community* with an interest in the fishery resources of the region, to prevent unacceptable impacts of such fishing practices...
>
> b) Take immediate action to reduce progressively large-scale pelagic driftnet fishing activities in the South Pacific region leading to the cessation of such activities by no later than 1 July 1991, as an interim measure, until appropriate conservation and management arrangements for South Pacific albacore tuna resources are entered into by the parties concerned;
>
> c) An immediate cessation to further expansion of large-scale pelagic driftnet fishing on the high seas of the North pacific and all the other high seas outside the Pacific Ocean, *with the understanding that this measure will be reviewed subject to the conditions in paragraph 4 (a) above...*[64]

This resolution is a typical outcome of the global environmental negotiating process.[65] For reasons which will be discussed in much greater detail in the final conclusion of this book, global environmental treaties have been characterized by inherent structural limitations designed to protect the perceived sovereign interests of the contracting states. Hence, the efforts to address environmental problems are not comprehensive. The scientific issues raised by the continuing degradation of the environment are not mirrored in the agreements which purport to be significant improvements in the ongoing negotiation process to improve the planet's biological health. Rather, the agreements themselves are characterized by least common denominator variables that are "easiest" to agree upon. In other words, a "solution" as such is not achieved. The process usually does not even come close. What is "achieved" is what can be agreed upon politically. This does nothing, of course, to address the underlying problems which exist quite independently from the machinations and activities of diplomats and politicians.

The 1989 resolution is no different. It is progress, albeit, in the sense the international community has declared that the issue is one which has "disturbed" many countries.[66] It is activity as usual—reflecting, of course, in all fairness the reality of the international system today—in that the resolution "recommends" a moratorium, but exceptions can be made when "effective conservation and management measures [can] be taken based upon statistically sound analysis ... jointly made by concerned parties of the international community" (see above). Who is to decide, and under what parameters is the decision to be subjected? Nothing in the resolution addresses this issue.

Greenpeace noted this weakness. It "welcomed the resolution as a 'step in the right direction,'" but observed that it failed to ask for an immediate halt to driftnetting in the South Pacific [and elsewhere]... Because the UN resolution leaves the door open for this destructive fishery to continue, ... [we are] going to take action to stop it."[67]

As Feld and Jordan noted in their 1988 study on international organizations, it is very difficult to measure the impact of INGOs on the decision-making process. This holds equally true for examinations of decision-making in national legislatures. I will address this issue in greater detail in the conclusion of this chapter. However, it is important to note that most members of the U.S. House of Representatives and Senate are certainly not unaware of the importance environmental issues in general have assumed, and specifically of the membership figures of an organization like Greenpeace; and they certainly pay attention to such indicators of public opinion as letters to their offices and public opinion polls taken by reputable organizations on environmental issues. Nonetheless this is not to say a clear line exists in tracing such influence. That task is better left to a separate study.

Most of the political activity of Greenpeace is generated at the level of the nation-state, operates in that realm, and seeks to influence the global system primarily through the national structures and institutions available through member countries. Two areas which demonstrate an exception to this general finding are Greenpeace efforts at the United Nations regarding an international ban on driftnetting, and attempts by Greenpeace International[68] to influence the London Dumping Convention.

The London Dumping Convention (LDC)—the official designation is the Convention on the Prevention of Marine Pollution by Dumping of Wastes and Other Matter—was adopted in 1972 and came into force in 1975. It is associated with the International Maritime Organization (IMO), which as a specialized agency of the United Nations is "... responsible for technical matters affecting shipping. Its primary task is to promote the safety of shipping by introducing standards which are internationally acceptable and implemented by many nations. A secondary function—and one which has grown in importance over the years—is the prevention of marine pollution from ships."[69] It is the second function to which Greenpeace International devotes its primary attention. The remarks which follow are more in the nature of a prolegomenon; they are not offered as definitive conclusions. Rather, what follows is more in the nature of a call for future research—research which will investigate the role words play in the formulation of policy statements. This etiological concern is assuming more importance as the relationship between INGOs and the international system undergoes significant change.

The LDC "... was adopted at a conference called by the United Kingdom, but [the] IMO has been responsible for duties associated with it since it came into force in 1975. The aim of the ... [instrument] is to prevent or limit the deliberate disposal at sea of various types of waste materials produced on land."[70] In providing secretariat functions for the administration of the convention, the IMO is the intergovernmental organization responsible on a day-to-day basis for the implementation and administration of this agreement. The organization in its "Strategy for the Protection of the Marine Environment" states that the convention itself

> is the global instrument for the control of marine pollution... The convention covers all marine waters seawards beyond the baseline and it requests each Party to take in its territory appropriate measures to prevent and punish conduct in contravention of the provisions of the Convention. The Convention prohibits the disposal of certain substances and wastes which, at the time the Convention was prepared, were felt to be particularly harmful, such as organohalogen compounds, mercury, cadmium, floating plastics, mineral oils and high-level radioactive wastes. Its contains specific regulations concerning the dumping of other materials which present a risk to the marine environment and human health. In addition, the Convention

> controls the incineration at sea of land-generated wastes and sets out criteria for the selection of dumping and incineration sites at sea.[71]

This strategy enunciated by the IMO has been criticized on grounds of weak or ineffective implementation. Greenpeace, in a policy statement released in April 1990 regarding expansion of the principles underlying the London Dumping Convention, observed:

> International agreements all too often reach decisions that reflect the lowest common denominator of compromise. In the case of the LDC, a Consultative Meeting is a pressurized market-place that often elevates the semblance of achievement above actual achievement.[72]

This problem, the substitution of reality for obfuscation and hope, is of course as old as the human condition and is not unique to diplomacy.[73] Contini and Sand, in the context of 20th century diplomatic problems associated with ocean degradation, have raised the issue

> whether traditional treaty techniques will prove to be suitable for meeting the technical requirements of effective 'ecomanagement' on the global or regional scale, once international action passes from the declaratory to the operational stage. Environmental problems characteristically require expeditious and flexible solutions, subject to current updating and amendments to meet rapidly changing situations and scientific/technological progress. In contrast, the classical procedures of multilateral treaty making, treaty acceptance and treaty amendment are notoriously slow and cumbersome.[74]

National governments have not recognized, in practice, this change in the international system. Patterns of interaction at the level of the nation-state have remained fundamentally *unchanged* since they were first established in the Renaissance. The resulting historical "methodology" of diplomacy, therefore, of states negotiating and attempting to address their problems primarily through their respective foreign offices, resident diplomats and (particularly in the post-Second World War era) the increasing utilization of large international conferences has remained the rule. Because of this, nation-states usually propose their measures for environmental protection and reform, including specific ocean degradation issues, within the established paradigm. One such recent proposal has concerned a debate over the proper wording for guidelines applying to national governments in their efforts to preserve the remaining marine ecosystems.

Greenpeace International has both embraced and suggested to the LDC Steering Group that *The Precautionary Principle*, a proposal first advanced by the Federal Republic of Germany in 1989, not only be adopted but further strengthened by embracing what could be called an "Ultra-Precautionary Principle." The Precautionary Principle itself is given below.

> The Contracting Parties accept the principle of anticipatory protection to be a framework for safeguarding the marine ecosystem by elimination of dumping of substances that are persistent, toxic and liable to bioaccumulate. This applies especially when there is reason to assume that certain damage or harmful effects on the living resources of the sea are likely to be caused by such substances, even when there is no scientific evidence to prove a causal link between dumping on the one hand and severe effects on the other hand (the principle of precautionary action).[75]

Greenpeace International believes this a positive step, but that it does not go nearly far enough. They have argued for a more rigorous and extensive protection. It is again useful to quote at length from the Policy Principles paper submitted in April 1990.

> Although these definitions for implementing the Precautionary Principle represent a step in the right direction, they do not entirely endorse a true precautionary approach. [1] Indeed, a decision to prevent and eliminate inputs of substances which possess all three characteristics of toxic, persistent and bioaccumulative, is based more on obvious common sense than on a precautionary approach. [2] Elimination of such substances alone, without more, will not provide the final solution to protection and restoration of marine ecosystems.[76]

This division of the problem, between an (insufficient) Precautionary Principle and the alternative (an Ultra-Precautionary Principle) is, however, unclear and confused on several points. Point "1" above denotes two components of the sentence: prevention of the introduction of dangerous substances into the marine environment, and in point "2" the insufficiency of "elimination" as an approach to problem solving.

First, let me state what it is that Greenpeace International favors by way of adoption as a "guiding force" in shaping LDC decision-making on this issue. In October 1989 the Nordic Council's International Parliamentarian Conference on the Pollution of the Seas met in Copenhagen and adopted the following statement. This declaration, Greenpeace contends, provides an answer to the problem of insufficient scope contained within the principle originally advanced by the Federal Republic of Germany. It notes

> The need for an effective precautionary approach, with that important principle intended to safeguard the marine ecosystem by, amongst other things, eliminating and preventing pollution emissions where there is reason to believe that damage or harmful effects are likely to be caused, even when there is inadequate or inconclusive scientific evidence to prove a causal link between emissions and effects.

The Policy Paper immediately follows with this explanation.

> This definition illustrates the role that a truly precautionary approach should play in the protection of the marine ecosystems by including all potentially damaging substances within the scope of application. It provides the level of protection necessary to restore and safeguard the marine ecosystems from further disruption caused by pollution.[77]

Sections of this Greenpeace International LDC statement are easy to comprehend. Others are not. A few examples will suffice.

The use of the word "dumping" offered by the Federal Republic of Germany would appear to be a key word in the definition. Opposition to procedures which result in either the manufacture or distribution and disposal of substances that are toxic, persistent and liable to bioaccumulate are easy to understand. It is also easy to comprehend that Greenpeace International would wish to go beyond the strict causal links some might argue must be present before implementing mandatory prohibitions. One might reasonably have very serious (and accurate) concerns with regard to the methods employed by industry and the nation-state before strict scientific experiments have demonstrated conclusively the danger posed by certain substances. The German definition hinges on the word "dumping." Should the substances be allowed (produced, stored and distributed), even before the question of what to do with them comes into play? Such appears to be the case.

The report states, however, that the "[e]limination of such substances alone, without more, will not provide the final solution to protection and restoration of marine ecosystems."[78] Does this mean Greenpeace is opposing the manufacture of the substances? On the one hand, this would certainly appear to be the case. Banning even the production of such material would go a long way to effectively eliminating the concerns expressed over what to do with the substances once created. If this be the case, why not simply say so? The issue also appears, however, to hinge on the degree to which the Contracting Parties would actually integrate the principle of "anticipatory protection" in a plan to effectively eliminate the introduction of polluting substances into the marine ecosystem. Is the admission of the necessity for this protection more a function of the international system agreeing to defer real action by substituting declaratory judgments as opposed to operational procedures for correction as Contini and Sand contend so often occurs? Or, does the prejudice expressed for the legitimacy of this "anticipatory protection" signal a serious commitment to reverse the pollution problem? Both enquiries would move this research beyond the confines of that intended by the present study. However, it does not appear to be a clear question just what it is Greenpeace is trying to persuade the contracting parties to undertake.

The matter is further complicated when the statement goes on to note

> Greenpeace International strongly recommends that the Precautionary Principle, if it is to reflect a true change in favour of a more realistic solution, be adopted as a broad guiding policy pursuant to which concrete measures to prevent the generation of wastes should be agreed and implemented.[79]

Is not what is trying to be said that the substances should not be produced in the first place?[80] Is it not the "generation" of wastes, but the "identification" of their simple yet critical existence, combined with their probable consequences to the marine environment, that is the the real cause for concern?

Conclusions

Paradoxically, but not overly so when one considers the evidence presented above, INGOs exhibit characteristics which are both weak and strong. They are able, at times, to prod governments into actions which otherwise they often would prefer not doing (at least within the time frame and referent points the INGOs seek, i.e. a faster implementation mode with more wide-ranging and scientifically effective measures than the political compromises state authorities would wish to implement), and they can bring to the attention of these same state authorities problems and issues which heretofore were indistinct or ill-defined. INGOs have a strength in being transnational in their scope and character, and the limited successes they have achieved (for example, Greenpeace with the tuna-dolphin issue and American tuna canners) is directly related to the ability of these groups to publicize their research findings and mobilize public opinion. The importance of such a limited role should not be underestimated, even if the "wins" are isolated events achieved after overwhelming efforts have been made. Parenthetically one may note the danger for those who are philosophically committed to supporting a change in the authority structure of the current state system (based, as it is, on sovereignty and territoriality) to overstate the power and influence of INGOs, and hence their influence in the contemporary international system.

As was mentioned in the Introduction to this chapter, the numbers of these organizations have certainly grown considerably since the end of the Second World War. Sheer size, however, does not guarantee significant authority or strength.[81] The state system is very strong, and INGOs have only made temporary advances on the problems they have addressed. These tactical movements of an actor within the state dominated configuration of world politics are subsumed, thereby, under the over-arching systemic hold or mastery of the sovereign state. This is not to say such an organizational pattern of world politics will endure indefinitely. It does say, however, that the most important player in this very important game of global environmental politics is still the sovereign state.[82]

Chapter 5 Notes

[1]For reasons which will be explained below, I define the protection of cetaceans—specifically dolphins—to be an issue of concern to the health of the overall marine environment.

[2]See Table 6.1 in Werner J. Feld and Robert S. Jordan, with Leon Hurwitz, *International Organizations, A Comparative Approach*, 2nd ed. (New York: Praeger Publishers, 1988), p. 212. In addition, see Robert S. Jordan, "Interest-Group Politics in Global Conferencing: An Assessment of the Contemporary Role of Nongovernmental Organizations in Intergovernmental Organizations," *SICA Occasional Papers Series, Second Series, No. 5*, Section on International and Comparative Administration, American Society for Public Administration. Portions of this chapter draw heavily upon the former.

[3]Werner J. Feld and Robert S. Jordan, op. cit., p. 24.

[4]For a biographical look at the Club and its founder, see Gunter A. Pauli, *Crusader For The Future, A Portrait of Aurelio Peccei, Founder of the Club of Rome* (Oxford: Pergamon Press, 1987). See also Alexander King, "The Club of Rome and Its Policy Impact," in William M. Evan, ed., *Knowledge and Power in a Global Society* (Beverly Hills, CA: SAGE Publications, Inc., 1981).

[5]See the works by Donella and Dennis Meadows, et al., *Limits to Growth* (First Report to the Club of Rome), (Washington, DC: Potomac Associates, 1972) and Mihajlo Mesarovic and Eduard Pestel, *Mankind at the Turning Point* (Second Report to the Club of Rome), (New York: E. P. Dutton and Reader's Digest Press, 1974). Alexander King in his article "The Club of Rome and Its Policy Impact," op. cit., pp.. 212-213, noted the following in regard to the Meadows' work: "To sum up on *Limits to Growth*, it seems to me that it was extremely important for three reasons. First, it has triggered off a debate throughout the world which was

extremely timely and necessary... A second useful impact ... is a new awareness of the importance of the interaction of different trends and different sectoral policies... Finally, [it] is important as the first serious attempt to break into a new field of research, that of understanding the quantitative workings of the complex world system." Such commentary needs little additional exposition.

As the first INGO to obtain a widespread hearing on the findings of its policy studies, the Club laid the groundwork for public acknowledgment of subsequent global modelling efforts by other INGOs and scholars. Such global modelling, even twenty years after the original efforts, are still controversial; in large part because of the sheer complexity necessary for the accurate construction of sophisticated and nuanced mathematical models.

[6]Alexander King "The Club of Rome and Its Policy Impact," op. cit., p. 222.

[7]Werner J. Feld and Robert S. Jordan, with Leon Hurwitz, *International Organizations, A Comparative Approach*, 2nd ed. (New York: Praeger Publishers, 1988), p. 231.

[8]See Werner J. Feld and Robert S. Jordan, with Leon Hurwitz, *International Organizations, A Comparative Approach*, 2nd ed., op. cit., pp.. 223-230

[9]Perhaps neither of these questions are of much importance, however. After all, many species have become extinct in the tens of thousands of eons since biological life first arose on this planet. Notwithstanding the past record, the human race is assaulting biological life on its planet to an unprecedented degree. Is it rational to continue this to the point of bio-failure? What role, directly or indirectly, do cetaceans play in this ongoing process?

[10]Feld and Jordan, op. cit., p. 236.

[11]Ibid., p. 235.

[12]This section is drawn from Ibid., p. 236.

[13]See Diana Crane, "Alternative Models of ISPAs" in William M. Evan, ed., *Knowledge and Power in a Global Society*, op. cit., p. 37.

[14]Feld and Jordan, op. cit. p. 238.

[15]Richard Munson, *Cousteau, The Captain and His World, A Personal Portrait* (New York: William Morrow and Company, Inc., 1989), p. 268 notes that by the early months of 1989—the fifteenth anniversary of the society's founding—membership levels reached almost 246,000. The French counterpart to The Cousteau Society is Foundation Cousteau, with headquarters in Paris. The same year 60,000 individuals and families were dues-paying members. This information was confirmed in a telephone interview with Cousteau Society officials in August 1990.

[16]See testimony of Jean-Michel Cousteau before the Committee on Merchant Marine and Fisheries, U.S. House of Representatives, "National Ocean Policy Commission Bill–HR 1171, May 1987." (Provided by The Cousteau Society, Los Angeles, CA.) In addition, see "Statement by JMC [Jean-Michel Cousteau] for the

California State Lands Commission, Hearing [on] 12 April 1989," Los Angeles: The Cousteau Society, 1989. For background on the Antarctic issue as a primary focus for The Cousteau Society see "Cousteau's Washington Plunge, Charming the Town For Antarctica." *The Washington Post*, 21 September 1989, p. C2; and "French and Australians Kill Accord on Antarctic," *The New York Times*, 25 September 1989, p. A8. For earlier efforts of The Cousteau Society in regard to ocean policy see "The Cousteau Society, Ocean Policy Papers, May 1979," Los Angeles: The Cousteau Society (n.d.).

[17]This paragraph and the following is drawn, in part from a telephone interview with Neal Shapiro, an Environmental Analyst with the Cousteau Society in Los Angeles, C A (13 November 1989).

[18]The elder Cousteau, however, will not live forever; as he himself is fully aware. This is realized, of course, by the society. It is doubtful whether Cousteau's son, Jean-Michel, can achieve the fame, recognition and respect which has accrued to his father. To note this, nonetheless, is somewhat trite. Naturally, such influence of the older Cousteau, however amorphous, is not easily transferred. The critical point I wish to make here is that all these issues are fully appreciated by the society; and its continued existence is not dependent upon the immortality of the founder.

[19]This is not the place in which to examine the history of the elder Cousteau and the transformation of his efforts as a young man from one who was simply interested in the sea and its mysteries to being the "persona" he is today. Cousteau is also a co-inventor of useful oceanographic equipment: the scuba or aqua-lung, deep-sea cameras, small submarines for ocean research (the "Diving Saucer"), marine science laboratories housed in floating buoys, undersea habitats, and the "Turbosail"—a wind propulsion system. His wide-spread recognition as a media star developed in the late 1960s. These stories of The Cousteau Society and Cousteau himself have already been told. The inventions are listed in a brochure of The Cousteau Society provided to the author by the society's offices in Los Angeles. In addition, see Richard Munson, *Cousteau, The Captain and His World, A Personal Portrait,* op. cit.; and Axel Madsen, *Cousteau, An Unauthorized Biography* (New York: Beaufort Books Publishers, 1986).

[20]Even before Jacques-Yves Cousteau become well known to the public at large as an advocate of ocean protection and integrity (beginning in the late 1950s), others also were raising cries of alarm at the threat posed to the water planet. See Thor Heyerdahl, *Kon-Tiki, Across the Pacific by Raft* (Chicago, New York and San Francisco: Rand McNally & Company, 1950).

[21]Richard Munson has reviewed the charges. He has noted: "The Captain and his staff have worked hard to promote and protect the Cousteau legend. Not a modest man, he has cultivated an image of himself as 'the world's greatest explorer,' the ingenious inventor of undersea equipment, and the ocean's best champion. But scores of colleagues and experts see things differently. Cousteau,

witness,' that is, accepting responsibility for being aware of an injustice. The Greenpeace ethic is not only to bear witness, but also to take action, direct nonviolent action, to stop atrocities against the environment. Thus we have used highspeed, inflatable craft to interfere with whalers and sealers, scaled smokestacks of air polluters and plugged discharge pipes of chemical polluters. We seek to obstruct wrongs without committing violence ourselves." See pamphlet "Greenpeace International," op. cit., and membership brochure, no title (Washington, DC: Greenpeace, n.d.).

[31]The qualifications for employment are of a high caliber, although commitment to the group's "cause" is, of course, an important component. See the advertisements for an International Project Coordinator and three Regional Project Coordinators for the Pulp and Paper Project in *New Scientist*, 21 July 1990, p. 90. The stated responsibilities for the positions include, respectively: "The successful applicant would assume responsibility for developing the on-going Pulp and Paper Project. Greenpeace opposes the use of chlorine bleaching by the paper industry and promotes alternative processes compatible with the nature and sustainable systems of forest management. The post is responsible for developing the Project's specific goals and strategies and their implementation internationally. Other duties include supervising a campaign team and administering a six-figure budget." And, "[t]he Pulp and Paper project also requires a Regional Coordinator to work in Europe. Regional Coordinators are required to work on the International Pesticides project in Europe and one, based in New Zealand, for [the] South Pacific. The project is directed at ending the production and trade of selected hazardous pesticides and promoting ecologically sound, socially just and economically viable agriculture. A Regional Coordinator, by delegation from the International Project Coordinator, supervises and develops the campaign in specific geographic regions. This involves working with national campaigners to plan and monitor research, non-violent direct action, public education and lobbying." The reader should note that the run-off from the industrial processes associated with the pulp and paper industry is a significant source of the land-based pollution that adversely impacts the global ocean.

[32]Greenpeace also has other offices in the United States located in the following cities: Chicago, Boston (Jamaica Plain, MA), Seattle, San Francisco and Fort Lauderdale (Wilton Manors, FL). By way of ancillary interest, Canadian offices are located in Vancouver, Toronto, and Montreal.

[33]This section is drawn from "Background Paper on the Eastern Tropical Pacific Tuna/Dolphin Issue, April 1988," pp. 3-8, prepared by Lesley Scheele and Dean Wilkinson of Greenpeace International, Amsterdam, The Netherlands. The paper itself was provided to the author by Greenpeace USA, headquartered in Washington, DC.

[34]As an example in 1959 the authors found an annual mortality for dolphins for U.S. and non-U.S. tuna fishing fleets to reach 55,000. The next year, 1960, the figure increased to 478,000. In 1961 the mortality figure was 534,000. In 1962 the figure dropped by 216,000 although it began again to increase throughout the 1960s and early 1970s. See "Background Paper on the Eastern Tropical Pacific Tuna/Dolphin Issue, April 1988," op. cit., Appendix-Table 1.

[35]See for membership of the Inter-American Tropical Tuna Commission (IATTC) the *1989-1990, Congressional Directory, 101st Congress* (Washington, DC: U.S. Government Printing Office, 1989), pp.. 889-890. The statistics from the IATTC are quoted in Lesley Scheele and Dean Wilkinson, "Background Paper on the Eastern Tropical Pacific Tuna/Dolphin Issue, April 1988," op. cit., p. 4-5.

[36]See Scheele and Wilkinson, op. cit., pp. 5-6.

[37]Here I am referring only to those boats registered within the United States.

[38]Lesley Scheele and Dean Wilkinson, op. cit., pp. 6-7.

[39]Ibid., p. 9.

[40]"Mexican registrations now dominate the fleet with 43 vessels. The other vessels setting on dolphins in the ETP are from Venezuela (15), Ecuador (4), Vanuatu (4), Spain (2), and the Cayman Islands, Costa Rica, El Salvador, and Panama with one vessel each." National Marine and Fisheries Service (1987), *Environmental Assessment on the Proposed Modification to the Marine Mammal Regulations Regarding the Importation of Yellowfin Tuna Caught with Purse Seines in the ETP*. Quoted in Lesley Scheele and Dean Wilkinson, op. cit., p. 9.

[41]This is a primary reason why the overall position of dolphin protection is weak. Whatever rules and regulations are adopted in the United States for the American market, albeit the largest in the world, including tuna; the majority of the tuna fleet operating in the Eastern Tropical Pacific is foreign owned. Thus, if nothing else is done other than to increase and tighten regulations or prohibitions on the American boats, the effect on the policy issue as a whole will be minimal.

[42]Lesley Scheele and Dean Wilkinson, op. cit., pp. 10-11.

[43]Ibid., pp. 9-31.

[44]Some of these groups include the American Cetacean Society, American Humane Association, California Marine Mammal Center, Defenders of Wildlife, Friends of the Earth, The Humane Society of the United States, National Audubon Society, Oceanic Society and The Whale Center–some large, and others small organizations. See Lesley Scheele and Dean Wilkinson, op. cit., p. 1.

[45]Ibid., p. 32.

[46]As one example see the 20 February 1989 letter of Lesley Scheele, Small Cetacean Project Coordinator, Greenpeace International, sent from the Wilton Manors, Florida office to E.C. Fullerton, Regional Director, Southwest Fisheries Center, National Marine Fisheries Service, National Oceanographic and Atmospheric Administration in Terminal Island, California. This letter is a nine

page comment on the interim final rule published in the Federal Register (6 January 1989) "... on rules and regulations governing the Taking and Importing of Marine Mammals" (p. 1). Provided to the author by Greenpeace USA, Washington, DC.

[47]A partial listing of the references contained in the study are as follows: Administrative Law Hearings in 1977 and 1980, Final Decisions and a Final Rule promulgated in the Federal Register in 1977 and 1980, respectively; Annual Reports from 1981 through 1986 of the Inter-American Tropical Tuna Commission, several academic reports including A. Myrick, et al., "1987 Hypocalcemia in Spotted Dolphins chased and Captured by a Purse Seiner in the Eastern Tropical Pacific and National Marine and Fisheries Service reports, regulations and environmental assessments. Figures for 1987 through 1989 were supplied in a supplementary table supplied by Greenpeace, Fort Lauderdale, Florida.

[48]Telephone interview 5 October 1990 with Craig Sinclair, Greenpeace, Fort Lauderdale, Florida.

[49]See Testimony of Vance Hughes before the Subcommittee on Fisheries and Wildlife, Conservation and the Environment, Committee on Merchant Marine and Fisheries, U.S. House of Representatives, "For a Hearing to Address the Implementation of the 1988 Amendments to the Marine Mammal Protection Act and H.R. 2926 and H.R. 2948," 4 October 1989. In addition, Senator Joseph Biden also introduced legislation in the Senate to accomplish similar purposes. See S. 2044 (101st Congress, 2nd Session–January 1990). Biden's bill is entitled the "Dolphin Protection Consumer Information Act of 1990" and also requires labelling with information on the issue. The relevant notifications are: "The Tuna in this Product was Caught with Methods that Kill Dolphins," or "Dolphin Safe." The geographic area of application also pertains to the Eastern Tropical Pacific.

[50]In other words, even if these bills were to become law the bureaucracies, in this case the Department of Commerce, have to draw up the rules and regulations to execute the congressional mandate. This is published, of course, in *The Federal Register*. Other laws have been passed to protect cetaceans, primarily the 1972 Marine Mammal Protection Act. This law and subsequent amendments have proved to be grossly inadequate. Congress intended through this legislation to protect dolphins, whales and other marine mammals. The record since 1972 has not been hopeful. The decision by the tuna canning companies in the United States in April 1990 is the most important step taken to specifically protect dolphins, save perhaps for the 1972 law itself. There lies the irony. Congressional intent was largely thwarted for 18 years by the bureaucratic lethargy and opposition of various administrations and the political echelon of the Commerce Department. Finally, three corporations decide to buy only "dolphin safe" tuna and—if the decision is fully and faithfully implemented—the chances for survival of many species of dolphin will go up dramatically.

[51]See *The New York Times*, 16 April 1990, "'Epic Debate' Led to Heinz Tuna Plan."

[52]The other "competitors" of Heinz/Starkist were Van Camp Seafood Company and Bumble Bee Seafoods, Inc. On 12 April 1990 they, along with Heinz, announced their decision (the latter two following within hours of the Heinz decision) to buy only "dolphin safe" tuna. These three companies dominate the canned tuna market in the United States with 70% of all sales. See *The New York Times*, 12 April 1990.

[53]See *The New York Times*, 16 April 1990.

[54]Traci Romine of Greenpeace USA noted that "these developments [were] 'a major victory—the most important event since the passage of the Marine Mammal Protection Act of 1972." She "... cautioned, however, that much work needs to be done... 'We have to find a way to verify dolphin-safe fishing practices from ship to shelf, persuade big and small firms here and overseas to adopt a strict policy, and pass the tuna labeling legislation ... before Congress.'" See *Greenpeace*, July/August 1990, p. 23. Romine also noted in a telephone interview with the author on 8 October 1990 that for Greenpeace, at least, the question was not simply one of a "boycott." That would be just "... one personal action [in deciding not to buy cans of tuna. We wanted to try] empowering people [to help them achieve a more comprehensive] solution." In light of this many tactics were pursued. The educational efforts by Greenpeace and its collaborators assumed a primary importance. Rather than "simply boycotting," the organization sought to encourage individuals to contact Heinz and the other companies letting them know of their opposition as well as to call for corporate policies which only allowed purchases of acceptable or "dolphin safe" tuna. Greenpeace also purchased $1,000 worth of stock in the H.J. Heinz corporation and worked with private individuals, also stockholders, to pressure the company to adopt the favored stance.

[55]See *Greenpeace* (a bimonthly), September/October 1990, pp. 23-24.

[56]See Greenpeace videotape film "Driftnetting" of 29 June 1990. This VHS tape was shown to interested members of the U.S. Congress in June 1990. In addition, see "Earthtrust 1988 North Pacific Driftnet Expedition, Briefing Paper: High-Seas Wildlife Kills, Fisheries Depletion." Honolulu, Hawaii: Earthtrust, 1988.

[57]For a systematic classification of these life forms see Peter G.H. Evans, *The Natural History of Whales and Dolphins* (London: Christopher Helm, 1987), pp. 32-75. Also, see John May, ed., *The Greenpeace Book of Dolphins* (New York: Sterling Publishing Company, 1990), pp. 14-23, 152; and "The Human Threat, Hunting and Fishing," pp. 86-107, particularly pp.. 100-103.

[58]*Greenpeace* (a bimonthly), September/October 1989, p. 20.

[59]See *Greenpeace* (a bimonthly), September/October 1990, p. 23 and notes made by the author from the film "Driftnetting." Dr. Brian Coffey was the marine biologist associated with the project and in charge of collating data.

[60]See film "Driftnetting," op. cit.

[61]See *Greenpeace* (a bimonthly), March/April 1990, p. 21.

[62]Telephone interview with Gerry Leape, ocean ecology lobbyist for Greenpeace USA, 26 October 1990.

[63]See A/44/225, "Large-scale pelagic driftnet fishing and its impacts on the living marine resources of the world's oceans and seas," adopted 22 December 1989 without a vote. This resolution was adopted on the report of the Second Committee of the General Assembly. For a text of the report see A/44/746/Add.7.

[64]See A/44/225 (emphasis added).

[65]A very important and critical exception may have been made had the efforts to implement a Law of the Sea treaty been successful. They were not. Every American president since Eisenhower supported diplomatic efforts to secure an agreement. Under the Reagan Administration in 1981 the United States withdrew its support for the concept of a Law of the Sea treaty. See Clyde Sanger, *Ordering the Oceans, the Making of the Law of the Sea* (Toronto and Buffalo: University of Toronto Press, 1987), especially pp. 197-199.

[66]See A/44/225.

[67]These comments were made, respectively, by Gerry Leape and Mike Hagler both of Greenpeace USA. Hagler's comments refer to the "direct action" which Greenpeace has pioneered. See *Greenpeace* (a bimonthly), March/April 1990, p. 21.

[68]Greenpeace International is headquartered in Amsterdam, The Netherlands.

[69]See "Focus on IMO, Preventing Marine Pollution" J/2866/Rev. 1 (London: International Maritime Organization), p. 6. For a brief description of the IMO see "Institutional Profiles: International Maritime Organization (IMO)," *International Environmental Affairs* (Summer 1990), pp. 270-272.

[70]See Focus on IMO, Preventing Marine Pollution" J/2866/Rev. 1 (London: International Maritime Organization), p. 8.

[71]See "Strategy for the Protection of the Marine Environment," (London: International Maritime Organization, 1988), p. 11.

[72]"Policy Principles and Options for the Expansion of the London Dumping Convention," Submitted by Greepeace International to the First Meeting of the Steering Group on a Long-Term Strategy for the London Dumping Convention, 17-20 April 1990 at the International Maritime Organization, London, England, p. 3. Provided to the author by Greenpeace USA, Washington, D.C. The Steering Group is an effort by the IMO to attempt an overall or broad look at the potential fundamental contributions the organization can take in improving the state of the global ocean. In that sense, it is a conscious decision to avert the gaze, so to speak,

temporarily away from the day-to-day concerns that seize organizations and top staff and policy individuals and focus it on a longer term perspective.

[73]For general background to these problems as well as very useful compendiums of diplomatic practice, both past and more recent, refer to Garrett Mattingly, *Renaissance Diplomacy* (New York: Dover Publications, Inc., 1988); and R. P. Barston, *Modern Diplomacy* (London and New York: Longman, 1988).

[74]Paul Contini and Peter H. Sand, "Methods to Expedite Environmental Protection: International Ecostandards." *American Journal of International Law* (1972), p. 38. The authors also note: "If international standards for environmental quality are to be set by diplomatic negotiations, the technical components of those standards may well be outdated by the time agreement is reached, and even more so by the time the agreement enters into force."

[75]London Dumping Convention, Report of the Twelfth Consultative Meeting, LDC 12/16,1 (December 1989), quoted in "Policy Principles and Options for the Expansion of the London Dumping Convention," Submitted by Greenpeace International ... (op. cit.), p. 4.

[76]Ibid., p. 4

[77]This paragraph, and the one above, are from "Policy Principles and Options for the Expansion of the London Dumping Convention, Submitted by Greenpeace International ..." op. cit., pp. 4-5.

[78]Ibid., p. 4.

[79]Ibid., p. 5.

[80]This lengthy section on Greenpeace International and the London Dumping Convention is interesting in several respects. Assuming I am correct in the identification of this unclear stated position, the issue raised reflects concerns beyond the immediate one in question. Principally, INGOs must be very clear in what it is they are trying to get national governments and intergovernmental organizations, either in tandem or sequence, to embrace and (eventually) to adopt.

[81]See Warner J. Feld and Robert S. Jordan, with Leon Hurwitz, *International Organizations, A Comparative Approach*, 2nd ed. (New York: Praeger Publishers, 1988), pp. 210-214.

[82]The sovereign state, however, is not all-powerful either. The role of multinational corporations has been addressed by many political scientists, particularly in the last two decades. See Chapter 1 of this book, "Intergovernmental Organizations (IGOs) and Multinational Corporations (MNCs)." See also David H. Blake and Robert S. Walters, *The Politics of Global Economic Relations*, 3rd ed. (Englewood Cliffs, NJ: Prentice-Hall, Inc., 1987), especially Chapter 4, "The Multinational Corporations: Challenge to the International System;" and Joan Edelman Spero, *The Politics of International Economic Relations* (New York: St. Martin's Press, 1977), especially Chapter 4, "The Multinational Corporation and the Absence of Management."

6

Conclusions

It certainly seems there is some article in the newspaper or video report on television each day which confirms the widely shared belief that the environment is in serious danger and decay.[1] Political scientists have a responsibility to examine such a sentiment, and to test its validity specifically in light of what the discipline can contribute by way of an in-depth study in the related issues of governance and the allocation of power. International relations specialists pursue a similar mandate, but focus their studies, more specifically, on the realm of the interactions at the state-to-state level and on such non-state actors that are relevant.

This book has centered the scope of its analysis on the international system and examined it in its broadest sense, taking into account the fundamental bases of its creation, operation and defects as they relate to the ecological problems of the Global Ocean. This was done for two reasons. First, all projects must be broken down into manageable components. This is the methodological rationale. Secondly, however, and the more important epistemological point, is that the Earth is a water planet. Water is the single most distinguishing feature or characteristic of this Earth, the third planet out from the "G-type" star we call our Sun.[2] The global ocean which has dominated the evolution of life on this planet, and the subsequent patterns of social organization which later developed, can be viewed as a paradigm; a paradigm which has received little attention, overall, from practitioners of the international relations discipline. This has resulted in an unfortunate tendency, still prevalent, of scholars to reinforce and give increased weight and attention to the study of those problems within the traditional milieu of the field: military power, and the perceived constraints on the conventional notions of sovereignty that are viewed as relevant, and the immutable dominance of the state system. The choice of such an alternative model in this study was deliberate. It allows for the search of alternative patterns or structures in governing and the possibility of better understanding new ways and procedures for confronting successfully the global environmental challenges of our own age.

This work began asking a very simple question: Can variously cooperating governments of sovereign states, in part operating through intergovernmental organizations, actually solve the problems of the Global Ocean? In other words, can the state system adapt quickly and adequately enough to meet the needs of an international system radically changing in terms of the set of challenges facing it? This book has only begun to answer such critical and long-needed questions.

When this study began four years ago as a doctoral dissertation within the department of political science at the University of New Orleans I had in mind a set of likely assumptions about the nature and causes of the ocean degradation problem that in many ways have held firm during the course of this project. To be sure, there have been some important modifications. Nevertheless, the basic problem I perceived before beginning this research has remained the same. And that problem is this: The entire world community is facing the most serious threat it has ever had to confront because the environment itself, and the ocean environment in particular, is under such serious assault that the most basic and fundamental questions relating to human survival and quality of life are raised. The threat is greater than any war that has been fought in the long history of humanity, and the steps that will be necessary to reverse the problem are more complex than any previous set of challenges human beings have ever faced. This complex of environmental degradation and decay over the last two centuries, but especially pronounced since the end of the Second World War, places in peril the premise upon which life as we have known it subsists.

While the basic presuppositions have remained valid throughout the conduct of this study, some important caveats must be added. First among these is the role of intergovernmental organizations (IGOs). These entities were viewed by the author as clear and viable alternatives to the three hundred year-old tradition of sovereign states. They still are. But the transition from one form of international organization (the territorially based sovereign state), however organized and perhaps later adopted or modified, to another that is global in its problem-solving tasks will be much more difficult than originally envisioned. There are a host of reasons for this. Beginning with the absence of any transnational taxing authority to impose policy choices and allocate resources on a global scale, to the lack of a permanent and standing multinational military force to secure and maintain global peace, and including a lack of any authoritative adjudicating body with the means to enforce its' decisions any major lessening of the monopoly of international influence currently wielded by sovereign states is a long way off.

International nongovernmental organizations (INGOs) play an equally curious role in the contemporary state system. They are even more the subjects of a domineering and often capricious international order. To this writer, INGOs were a category of highly unique, talented and inherently mobile structures, innovative and potentially intrinsically universal in vision, operation and purpose. All these descriptions still remain sound. Here too, however, some modification is in order. INGOs remain catalysts for developing new and original thinking on common problems confronting the whole of the international community. They are motivators, both for publics and governments. INGOs can also be effective pressure group organizations on an international scale. They can bring influence to bear on the governments of states and intergovernmental organizations arising from policy preferences and positions that are adamant or rigid in political insistence; interposing tension and a degree of possible or latent coercion that otherwise would not be present. The very important qualifying factor, however, is the limited and episodic nature in which INGOs operate. Many factors constrain their potentially greater influence. Chief among these are a gross lack of financial resources when compared with that available to states. INGOs have much less in the way of access to financial assets than even IGOs; and the latter have been chronically underfunded (especially the United Nations system) in the post-war period. This especially was true during the throes of the dying Cold War years of the 1980s. It is still true in the beginning years of the 1990s. Such components may change as the world moves closer to the next century. Other factors impinge as well, though—the small size of professional staffs, the lack of large numbers of active supporters when compared with the size of the institutions they criticize or (hesitantly) oppose, including the consequent "junior-partner" status (if that much) that INGOs assume within an international system dominated by Western states; and, ultimately, an authority structure operating at the international level which requires only territory as a prerequisite to full participation or membership instead of such possible alternatives as "values" or "issue components."[3]

The question of whether IGOs provide an important, truly distinguishable and needed service to the international community, or whether they simply duplicate programs and activities already undertaken on the state level, is an important issue.[4] It is also a problematic one. The intergovernmental organizations examined make unique contributions to the study of ocean degradation issues as well as to the more general responsibilities incumbent upon any discipline which must seek significant contributions in the advancement of knowledge. Do IGOs, however, in their programs of research and dissemination of findings involve themselves with questions that are addressed nowhere else? In other words, are the contributions unique and unduplicated by other organizations, particularly those funded and administered by states? The answer is a qualified no. The main purpose of the numerous IGO programs examined in this work is to provide information, knowledge and services to states that otherwise would not have access. The reasons are varied and include the inability of certain states, on a bilateral basis, to secure such assistance directly. By annual contributions to IGOs such as UNESCO, states which are economically and technologically more advanced can help fund programs of aid in marine science to "underdeveloped" countries. Such programs are, of course, possible to arrange on bilateral bases. For instance, the foreign aid programs of a particular state can obviously include marine science assistance to those states desiring and needing this particular form of help. The more advanced states, again economically and technologically, have found it advantageous to pursue both avenues of assistance—indirectly, through multilateral organizations and directly through bilateral programs.

On the other hand, one cannot argue *a posteriori* that such a situation is essential or required. It may, however, be both practical and necessary. IGOs have developed, and states have joined them, because states have found interests that are fulfilled in these particular venues which are both advantageous to the state involved and necessary in that the same goals could not be as efficaciously pursued in other locales. In other words, IGOs exist not because of any affection states have for multilateral negotiations or programs, but because states believe unique advantages can be won by membership and participation.

The question of sources is always an interesting problem which confronts any researcher. In this work, the question has been which sources to use. There is a veritable excess in primary source material from intergovernmental organizations. In the case of the IGOs examined in the preceding chapters, these institutions publish reports and statements which are frequent, detailed and at times voluminous. I could only focus on what, in reality, was a minuscule proportion of the material available. For those with the patience and interest to wade through the documents the rewards, however, are great. They detail an account of social organizations attempting to grapple with serious and profound issues of planetary importance. Whether they will be more successful at actually solving the problems than has been the case in the past is essentially out of their hands. Power, the power to avoid taking action as well as the power to address issues with the intention of solving them, in the contemporary international system is in the hands of states. And, of course, among the groupings of states some states have much more power than others. It is another issue entirely to examine how might a transformation of the international system actually come about. That has not been the task here. It is a project which is waiting to be undertaken, particularly in light of the severe environmental issues confronting the human species. Sovereign states are not capable of solving the issue of ocean degradation, either on their own or cooperating with others. The problem is so immense it dwarfs the capabilities, human or financial, of any single state. When states operate together in joint forums, through IGOs as one prominent example, the procedures which are available in the traditional repertoire of diplomacy are inadequate. Constraints of time (treaty negotiation is a long and cumbersome process) and the necessity of taking up one or perhaps at the most two incredibly complex issues of a related environmental question is a woefully inadequate and dangerous procedure to utilize for making progress in this area. How might sovereignty, as currently understood and accepted by international society, be reformed and the bases of its operation be expanded, in the realm of ocean protection is another vital topic awaiting examination.

Sources available from international nongovernmental organizations (INGOs) present a different problem to the researcher. There, the availability of primary source materials in no way equals what can be secured from IGOs. While this does not present insuperable obstacles for research into the work and activities INGOs, it does require the student to devote more time to the search and acquisition of acceptable material. INGOs which publish a relatively large volume of newsletters, magazines (whether for their membership or the general public at large) and or reports are, therefore, more likely to be the subject of serious investigation.

The contemporary state system is unable to successfully address the issues of ocean degradation. To be sure, as was pointed out previously, limited and very narrow progress is and has been made on selected issues. But that is the critical point. Given the structural constraints imposed by the operation of the existing conceptions of sovereignty, progress on the issue of ocean pollution will be restricted and constrained at best. Yet the extant state system is dominated by the very powers which are the major blocking points to the reform and renewal which is required if dangerous environmental policies are to be reversed. This is the extreme irony of the human condition in the final years of the 20th century. The Industrial Revolution paradigm itself, moreover, as the greatest single factor responsible for raising the living standards of billions of people, is also the single most important reason as to why the world community finds itself today in the discordant relationship with the living complex of institutions and peoples we call Planet Earth.

The past is often a significant sign post to future developments. If the last three hundred years are anything to go by, and this author clearly believes they are, states will resist strenuously any surrender of their privileges and prerogatives. Only a fundamental change in the current system, however, will secure an ocean environment healthy and sustainable for all the many life forms on this planet that are inextricably intertwined.

Chapter 6 Notes

[1]Printed on non-recycled paper, and filmed on a substance petroleum based.

[2]See Jay M. Pasachoff, *Astronomy: From the Earth to the Universe*, 3rd ed., (Philadelphia: Saunders College Publishing, 1987), p. 316.

[3]See Lynn H. Miller, *Global Order, Values and Power in International Politics*, 2nd ed., (Boulder, CO: Westview Press, 1990), especially Chapters 4 and 8, respectively, "Power and Values in World Society," and "The Closing of the World Frontier."

[4]I acknowledge with thanks the comments of Clive Archer on this point and for bringing this question to my attention.

Acronyms

COMAR	Major Interregional Project on Research and Training leading to the Integrated Management of Coastal Systems
COSPAR	Committee on Space Research
DMS	Division of Marine Sciences
ETP	Eastern Tropical Pacific
GELTSPAP	Group of Experts on Long-Term Scientific Policy and Planning
GEMS	Global Environment Monitoring System
GESAMP	Group of Experts on the Scientific Aspects of Marine Pollution
GIPME	Global Investigation of Pollution in the Marine Environment
HLW	High-level radioactive waste
IATTC	Inter-American Tropical Tuna Commission
IGO	Intergovernmental Organization
IGOSS	Integrated Global Ocean Station System
IOC	Intergovernmental Oceanographic Commission
ILW	Intermediate-level radioactive waste
IAEA	International Atomic Energy Agency
ICSU	International Council of Scientific Unions
IGY	International Geophysical Year
IGBP	International Geosphere-Biosphere Program
IMO	International Maritime Organization
INGO	International Nongovernmental Organization
ISPA	International Scientific and Professional Association
LDC	London Dumping Convention
LLW	Low-level radioactive waste
MAPMOPP	Marine Pollution Monitoring Pilot Project
MNC	Multinational Corporation
NMFS	National Marine Fisheries Service
TEMA	Training, Education and Mutual Assistance

UN	United Nations
UNCLOS	United Nations Conference on the Law of the Sea
UNEP	United Nations Environment Programme
UNESCO	United Nations Educational, Scientific and Cultural Organization
WCP	World Climate Program
WCRP	World Climate Research Program
WOCE	World Ocean Circulation Experiment
WMO	World Meteorological Organization
UNIDO	United Nations Industrial Development Organization

Select Bibliography

Andresen, Steinar and Willy Ostreng, eds. *International Resource Management: The Role of Science and Politics*. London and New York: Behhaven Press, 1989.

Borgese, Elizabeth Mann and Norton Ginsburg, eds., *Ocean Yearbook 6*. Chicago: University of Chicago Press, 1986.

Boulding, Kenneth. *Ecodynamics: A New Theory of Societal Evolution*. Beverly Hills, CA: Sage, 1981.

Brown, Lester R. *World Without Borders*. New York: Random House, 1972.

Bull, Hedley. *The Anarchical Society, A Study of Order in World Politics*. New York: Columbia University Press, 1977.

Bull, Hedley and Adam Watson, eds. *The Expansion of International Society*. Oxford: Clarendon Press, 1984.

Carr, E. H. *The Twenty Years' Crisis, 1919-1939, An Introduction to the Study of International Relations*. London: Macmillan & Co., Limited, 1940.

Carroll, John E., ed. *International Environmental Diplomacy, The management and resolution of transfrontier environmental problems*. Cambridge, New York and Melbourne: Cambridge University Press, 1988.

Clark, R. B. *Marine Pollution*. Oxford: Clarendon Press, 1986.

Claude, Inis L., Jr. *Power and International Relations*. New York: Random House, 1962.

Claude Inis L., Jr. *Swords into Plowshares, The Problems and Progress of International Organization*, 4th ed. New York: Random House, Inc., 1971.

Crawford, James. *The Creation of States in International Law*. Oxford: Clarendom Press, 1979.

Dahlberg, Kenneth A. *Environment and the Global Arena, Actors, Values, Policies and Futures*. Durham, NC: Duke University Press, 1985.

DeFries, Ruth S. and Thomas F. Malone, eds. *Global Change and Our Common Future, Papers from a Forum*. Washington, DC: National Academy Press, 1989.

DeLupis, Ingrid Detter. *International Law and the Independent State*, 2nd ed. Aldershot, England: Gower Publishing Co., Ltd. 1987.

Donelon, Michael. *The Reason of States, A Study in International Political Theory*. London: George Allen & Unwin, 1978.

Evan, William M., ed. *Knowledge and Power in a Global Society*. Beverly Hills, CA: SAGE Publications, Inc., 1981.

Evans, Peter G.H. *The Natural History of Whales and Dolphins*. London: Christopher Helm, 1987.

Falk, Richard, Friedrich Kratochwil and Saul H. Mendlovitz. *International Law: A Contemporary Perspective*. Boulder, CO: Westview Press, 1985.

Feld, Werner and Robert S. Jordan, with Leon Hurwitz. *International Organizations, A Comparative Approach*, 2nd ed. New York: Praeger Publishers, 1988.

Finkelstein, Lawrence S., ed. *Politics in the United Nations System*. Durham, NC and London: Duke University Press, 1988.

Galtung, Johan. *The True Worlds, A Transnational Perspective*. New York: The Free Press, 1980.

Gross, M. Grant. *Oceanography, A View of the Earth*, 4th ed. Englewood Cliffs, NJ: Prentice-Hall, Inc., 1987.

Haas, Ernst B. *Beyond the Nation-State, Functionalism and the International Organization*. Stanford: Stanford University Press, 1964.

Hardin, Garrett and John Baden, eds. *Managing the Commons*. San Francisco: W.H. Freeman, 1977.

Hargrove, John Lawrence. *Who Protects the Ocean?* St. Paul, MN: West Publishing Co., 1975.

Hinsley, F.H. *Sovereignty*, 2nd ed. Cambridge: Cambridge University Press, 1986.

Holsti, K.J. *International Politics, A Framework for Analysis*, 2nd ed. Englewood Cliffs, NJ: Prentice-Hall, Inc., 1972.

James, Alan. *Sovereign Statehood, The Basis of International Society*. London: Allen & Unwin, 1986.

Jacobson, Harold K. *Networks of Interdependence, International Organizations and the Global Political System*. New York: Alfred A. Knopf, 1979.

Kay, David A. and Eugene B. Skolnikoff, eds. *World Eco-Crisis, International Organizations in Response*. Madison and London: University of Wisconsin Press, 1972.

Keohane, Robert O. and Joseph S. Nye. *Power and Interdependence, World Politics in Transition*. Boston and Toronto: Little, Brown & Co., 1977.

Leonhard, H. Jeffrey. *Pollution and the Struggle for the World Product; Multinational Corporations, Environment and International Comparative Advantage*. New York: Cambridge University Press, 1988.

Lovelock, James. *The Ages of Gaia, A Biography of Our Living Earth*. New York: W.W. Norton & Co., Inc., 1988.

Madsen, Axel. *Cousteau, An Unauthorized Biography*. New York: Beaufort Books Publishers, 1986.

McLuhan, Marshall and Bruce R. Powers. *The Global Village, Transformations in World Life and Media in the 21st Century*. New York and Oxford: Oxford University Press, 1989.

Meadows, Donella, et al. *The Limits to Growth*. New York: Universe Books, 1972.

Mesarovic, Mihajlo and Eduard Pestel. *Mankind at the Turning Point, The Second Report to the Club of Rome*. New York: E.P. Dutton & Reader's Digest Press, 1974.

Miller, Lynn H. *Global Order, Values and Power in International Politics*, 2nd ed. Boulder, CO: Westview Press, 1990.

Morrall, John B. *Political Thought in Medieval Times*. London: Hutchinson & Co., Ltd. 1958. (Reprinted Medieval Academy of America, 1987.)

Munson, Richard. *Cousteau, The Captain and His World, A Personal Portrait*. New York: William Morrow and Company, Inc., 1989.

Nagel, Stuart S., ed. *Environmental Politics*. New York and London: Praeger Publishers, 1974.

Our Common Future, The World Commission on Environment and Development Oxford and New York: Oxford University Press, 1987.

Park, Chris C., ed. *Environmental Policies, An International Review*. London, Sydney and Dover, NH: Croom Helm, 1986.

Pauli, Gunter A. *Crusader For The Future, A Portrait of Aurelio Peccei, Founder of the Club of Rome*. Oxford: Pergamon Press, 1987.

Pirages, Dennis. *The New Context for International Relations: Global Ecopolitics*. North Scituate, MA: Duxbury Press, 1978.

Pirages, Dennis. *Global Technopolitics: The International Politics of Technology and Resources*. Pacific Grove, CA: Brooks/Cole Publishing Co., 1989.

Quigg, Philip W. *A Pole Apart, The Emerging Issue of Antarctica*. New York: McGraw Hill Book Company, 1983.

Richardson, Jacques G., ed. *Managing the Oceans; Resources, Research, Law*. Mt. Airy, MD: Lomond Publications, Inc., 1985.

Rittberger, Volker, ed. *Science and Technology in a Changing International Order, The United Nations Conference on Science and Technology for Development*. Boulder, CO: Westview Press, 1982.

Repetto, Robert, ed. *The Global Possible, Resources, Development and the Next Century*. New Haven and London: Yale University Press, 1985.

Sanger, Clyde. *Ordering the Oceans, The Making of the Law of the Sea* (Toronto and Buffalo: University of Toronto Press, 1987.

Sprout, Harold and Margaret Sprout. *Towards a Politics of the Planet Earth*. New York: Van Nostrand Reinhold Co., 1971.

Sterling, Richard W. *Macropolitics, International Relations in a Global Society.* New York: Alfred H. Knopf, Inc., 1974.

Taylor, Paul and A. J. R. Groom, eds. *International Institutions at Work.* London: Pinter Publishers, 1988.

Tree of Life, Buddhism and Protection of Nature. Buddhist Perception of Nature, 1987.

Walton, D. W. H., ed. *Antarctic Science.* Cambridge, New York and Melbourne: The Press Syndicate of the University of Cambridge, 1987.

Ward, Barbara and Rene Dubos. *Only One Earth.* New York: W. W. Norton & Co., Inc., 1972.

Wight, Martin. *Power Politics*, ed. by Hedley Bull and Carsten Hollbraad. New York: Holmes & Meier Publishers, Inc., 1978.

The following offices, journals or publications are useful in following environmental questions in general and ocean degradation issues in particular. Original source material for this book was drawn primarily from United Nations documents and reports or studies from international environmental groups such as Greenpeace and The Cousteau Society. Specific references to the material utilized can be found at the end of the various chapters under the "Notes" section.

Ambio
Calypso Log
Greenpeace
International Environmental Affairs
Science
The New York Times
Worldwatch Institute
United Nations (sources)
- Group of Experts on the Scientific Aspects of Marine Pollution (GESAMP)
- International Atomic Energy Agency (IAEA Bulletin)
- International Maritime Organization
- World Meteorological Organization
- United Nations Educational, Scientific and Cultural Organization
- United Nations Environment Programme
- Reports of the Secretary-General on the Protection and Preservation of the Marine Environment

Index

Herman Prager is Assistant Professor of International Relations at Northeast Louisiana University in Monroe. He received his Ph.D. in political science from the Unversity of New Orleans, and the undergraduate and post-graduate degrees from Loyola University (New Orleans) and the University of California, Santa Barbara.